STATEN ISLAND
AND THE
VERRAZZANO-NARROWS BRIDGE

STATEN ISLAND AND THE VERRAZZANO-NARROWS BRIDGE

CHAOS COMES TO A COMMUNITY

PATRICIA M. SALMON

THE History PRESS

Published by The History Press
An imprint of Arcadia Publishing
Charleston, SC
www.historypress.com

First published 2026

Manufactured in the United States

ISBN 9781467170321

Library of Congress Control Number: 2025945976

For my brothers Jim and Frank. We began the journey together.
Unfortunately, it did not continue.

CONTENTS

Acknowledgements 9
Introduction 11

1. What Preceded the Bridge 15
2. Pre-Bridge Industry on Staten Island 21
3. The First Real Estate Frenzy 24
4. Planning for the "Narrows Bridge" 36
5. Building the Bridge 43
6. The Opening of the "New Colossus" 58
7. Connecting Highways and Environmental Challenges 62
8. "Jerry-Built, Rubber Stamp Row Houses Spring Up in the Sylvan Wilderness" 72
9. Unplanned Chaos and Disorder Continues 84
10. No Maps, No Roads, No Sewers, No Schools—No Problem 89
11. The Bane of Many: Septic Tanks and Sewers (Or Lack Thereof) 103
12. The Population Explodes 108
13. Let's Do the Portable Shuffle 113
14. Urban Renewal on the South Shore? 141

15. Aftereffects: From Transportation to Traffic 148
16. Rising Crime 152
17. Healthcare Concerns for an Expanding Population 158
18. The Shopping Experience Evolves 164
19. Recreation, Staten Island Style 166
20. Preserving the Historic Past 171
21. Staten Island Responds 178
22. Uniformed Personnel, Utilities, and Other Services 184
23. Overdevelopment 192

In Conclusion 195
Notes 201
About the Author 219

ACKNOWLEDGEMENTS

Special thanks to Barbara Hemedinger, Dr. Cheryl Criaris-Bontales, Norma Su Wolfe, Dr. Gregory Adamo, Ciro Galeno, Dawn Daniels, Carlotta DeFillo, Carli DeFillo, Cliff Hagen, Bill Higgins, Tina Kaasmann-Dunn, Nick and Christie Pisacano, Marianna Randazzo, Giuseppe Settinieri, Joan Sidenberg, Jan Somma, the *Staten Island Advance*, Staten Island Historical Society, Staten Island Museum, and Steve Zaffarano.

The *Staten Island Advance* did an amazing job of covering the changes that occurred before, during, and after the opening of the Verrazzano-Narrows Bridge. I would like to acknowledge the following reporters: Mike Azzara, Philip Brittain, Kenneth Fanizzi, Drew Featherston, Victoria Flanders, William Franz, S.S. McSheehy, Mark Weisner, James Whitford, Raymond Wittek, and those whose names did not appear in the byline.

INTRODUCTION

With almost every part of New York a horrible example of the blight caused by plan-less development, it is criminal that this last chance for any approach to civic grace is being sacrificed on the altar of municipal and private greed.
—New York Times, *August 10, 1965*

November 21, 2024, marked the sixtieth anniversary of the opening of the Verrazzano-Narrows Bridge. The bridge spans the Narrows, that slim tract of open water between Fort Wadsworth, Staten Island, and Fort Hamilton, Brooklyn. While the opening of the Verrazzano-Narrows Bridge was celebrated by some, borough natives would refer to it as the most cursed, vile event to have ever occurred since permanent settlement in 1661. They believed that their bucolic borough was forever changed. They were right. Politicians would never be forgiven for allowing this blemish of a bridge to be built. But some of the most derisive comments would be aimed at the transplants who migrated across the bridge to make a new life on Staten Island. These people will forever be known as Brooklynites.

I am a Brooklynite. My father grew up amid the crowded urban conditions of Bedford-Stuyvesant, Brooklyn. This congestion left its mark. After he married my mother, the young couple moved to Bensonhurst and then to Flatlands. These were Brooklyn neighborhoods that had actual houses, as opposed to apartment buildings. But my father's need for open space was so great that he bought an isolated quarter-of-an-acre piece of property at Bulls Head, Staten Island. This resulted in the Salmon family relocating to the

"Forgotten Borough" in May 1962. We were nestled in between a farm to the north, an empty lot to the east, and the Merrell Homestead Graveyard to the west. Across the unpaved divide of Merrill Avenue were two more farms, small but still working farms.

The isolation would not, nor could it, last forever. That bridge was being built at the time of our move. It would conquer the Narrows and Staten Island. The end result was that transport between the two boroughs was now convenient. Other Brooklynites, many other Brooklynites, soon had the same idea as my father. The promise of the modern bridge encouraged and enticed thousands more to desert Brooklyn for the open acres of Staten Island. The masses were deserting King's County. Never mind Brooklyn's motto: "In unity, there is strength." These citizens were only interested in getting to the least populated borough of New York City. Surveyors were soon surveying, builders building, and truckers trucking. But first, those who had purchased these golden acres sold off the enriched topsoil of farms at Bulls Head, New Springville, Travis, Graniteville, Green Ridge, and other produce-producing communities. The topsoil would be resold to the highest bidder and spread to God only knows where. Two-family mother/daughter ranch houses were the new "produce" to sprout from this once abundant topsoil. These houses were coveted by the escaping Brooklyn transients. A whole second floor of bedrooms, along with a dining room, kitchen, and bath, would be accompanied by a family room and garage on the first floor. The three rooms adjacent to the garage on the ground floor would allow young families to put mom in the apartment below—hence the term *mother/daughter ranch*. Or the apartment could be rented, with the rent paying off the monthly mortgage. The island was changed forever.

With the opening of the bridge, breadwinning transplants now had an easy and supposedly fast commute from the outlying borough to their jobs in other boroughs. But such words as *rural* and *easy commute* would soon fall by the wayside. Much to the dismay of the recently arrived, and even more to the dismay of the natives, more and more Brooklynites fled to Staten Island.

The Salmon family move was wedged between breaking ground on the bridge and its ribbon cutting. We were not Staten Islanders, but we were not exactly the Brooklynites that the natives hated, either. Or were we? One former colleague informed me that my *Staten Island Advance* obituary will read "born in Brooklyn"—a statement on Staten Island that is more heartbreaking than the most inaccurate accusation.

As someone who grew up romping through the fields and woodlands of the island, and as someone who watched the unparalleled destruction of those natural areas take place, I am not only writing a history narrative. I am also trying to make sense of what occurred sixty years ago. It has been a daunting and difficult effort, one that could fill a dozen volumes on the subject. Many events, activities, lies, and betrayals occurred to make Staten Island what it is today.

I, as a child, could never have fully realized what was going on. All I knew was that Merrill Avenue was alive with large, loud, intrusive trucks of a size that would never be allowed on such a road today. At its best, Merrill Avenue was busted up and hazardous during the early to mid-1960s. If you traveled toward its intersection with South Avenue, it actually became a dirt road. Rest assured that there were many other such roads throughout the borough at this time. As the trucks, bulldozers, and cement mixers started traversing this upgraded path, I was very displeased at the fact that the woods, wetlands, and meadows that were once at my disposal were literally disappearing before my very own eyes. Focusing those young eyes, I soon saw rows of tract housing sprouting across the neighborhood.

In looking back and researching the events leading up to the opening of the Verrazzano-Narrows Bridge, I have concluded that those in power did little to prepare for what was going to occur on Staten Island. I hope they were oblivious, because a free-for-all ran rampant from one shore of Staten Island to the other. The bedlam that occurred was unplanned, chaotic, and disorganized. It did benefit the City of New York and select others. And yes, it did benefit the "Brooklynites," but they suffered as well.

So here is a far-too-brief look at the bridge and its immediate aftereffects on certain aspects of borough life. These changes occurred swiftly and blindly, and it left acrimony in its wake. And yes, we are still experiencing the problems that resulted from this era in Staten Island history.

Chapter 1

WHAT PRECEDED THE BRIDGE

Staten Island is the southernmost point of New York State. It is also the county of Richmond. The county was established on November 1, 1683, by Colonial Governor Thomas Dongan. It was one of the original ten counties in what was then the province of New York; residents voted to become part of the City of New York in 1898.

The county of Richmond was dubbed Staten Island by the Dutch, the original European claimants of this mass of land. Richmond and Staten Island are used interchangeably—officially and unofficially—and Staten Island is, unofficially, the more popular of the two. It is simply not true that a Dutch sailor, on entering what would become the New York Harbor, said, "Is that an island?" and these words were later corrupted by island residents to sound like "Staten Island." The name Staten Island is derived from the Dutch Staaten Eylandt. It was named the Island of the States in honor of the Estates General of the Netherlands, the seventeenth-century governing body of that nation.

The first permanent settlement at Staten Island took place at what is now Arrochar in 1661. The British took over in 1664, and for the next three hundred years the island remained rural, agricultural, and isolated. Of course, it is an island, so the only way to depart from or arrive on it was by boat—until 1889, when a railroad bridge was built between Elizabeth Town, New Jersey, and Old Place, Staten Island.

The earliest agriculturists on the island were the Lenape. They shared their knowledge of farming with the Europeans who settled there. These settlers consisted of Dutch, French Huguenot, and English farmers who owned enslaved people. Enslavement officially ended in 1827.

During the early days of farming, hay, corn, wheat, potatoes, oats, rye, barley, and buckwheat were vital crops. Later produce included tomatoes, berries, apples, and pears. Poultry, hogs, cattle, wood, and goods made by hand were important products that could be sold to the public when not needed by the family. Indeed, Staten Island's greatest market was always Manhattan, followed by Brooklyn and port locations in New Jersey, especially along the Arthur Kill River and the Kill Van Kull. Many of the early farmers were also oystermen who gathered this high-quality bivalve as well as clams from local waters.

Farming continued to be an important occupation on the island, although the continuous growth of industry after 1819 provided many blue-collar jobs. Dental goods, beer, bricks, terra cotta, soaps, bluing, enzymes, paint, handkerchiefs, candy, boats, and an assortment of other products were manufactured in the borough until the 1960s. By the 1950s, much of central, southern, and western Staten Island was still cultivated, mainly by Greek American and Italian American farmers.

Greek immigrant Anthony Chrampanis, who earned twenty-two dollars a month by rolling cigars at the American Tobacco Company in Manhattan in 1906, saved enough money to buy New Springville acreage—acreage that he would continually nourish and that would be highly prized before, during, and after the construction of the bridge. Eventually, the Chrampanis spread would grow to fifty acres. Celery, chicory, dandelions, escarole, leeks, parsley, scallions, and Swiss chard were commonly grown by the Greek farmers along with the previously mentioned crops. The money Anthony made allowed additional family members to leave the crushing poverty of Lemnos, Greece, to start a new life in the United States. The Chrampanis family was still farming twenty acres in the mid-1960s. Their flourishing roadside stand was popular until 2007. In the 1950s and onward, developers constantly knocked on their door looking to buy the farm. In 1972, the family sold their remaining property for tax reasons, although they did keep one acre for cultivation. On this plot, a greenhouse was installed. It was one of the last commercial farms in New York City. Today, PS 69 and the Laurie Intermediate School stand on land once tilled by the Chrampanis family.

Growing up in Bulls Head across from the small Merrill Avenue farm cultivated by Bill Criaris, his daughter Cheryl and I would sell five-pound bags of juicy red tomatoes on the brick stairsteps that fronted the ever-increasing automobile traffic of Bulls Head during the mid-1960s. Even the

Harvesting corn at the Chrampanis Farm, New Springville, 1937. *Left to right*: George Talumbos, Bill Criaris, Fritz Pandazis, Lou Chrampanis, and Nick Susalis. *Courtesy of Dr. Cheryl Criaris-Bontales.*

truckers who transported building materials to the developments springing up around Arlene Street would stop to make a purchase. I have no doubt that they were never disappointed. The tomatoes grown on the Criaris farm were the most delicious I have ever tasted in my life. Mr. Criaris also grew eggplant, zucchini, peppers, and more. Bill even installed a greenhouse on the farm in the 1990s to establish a small business selling plants to various concerns. Bill Criaris operated that farm from 1958 until a few months before his death on November 27, 1995.

Vegetable stands were common on the island. The setups were simple: Boards were laid across barrel tops, and the produce was placed atop the boards. A scale was the only thing needed to complete the enterprise, which sometimes led to a permanent business. These family businesses were established on the busiest road that bordered the farm.

Most nineteenth- and twentieth-century immigrant farmers hoped their children would not have to do backbreaking farmwork. They hoped their offspring would be diligent students who would become lawyers, doctors, or some other kind of professional. Their hopes were often realized thanks to the steady guidance of hardworking parents and grandparents.

As he watched the Con Edison power plant in Travis spewing smoke, Bert Bifulco lamented the effects of air pollution on his crops. The thirty-seven-

The Chrampanis family farmstand, Richmond Avenue, New Springville, 1937. Photo by William Criaris. *Courtesy of Dr. Cheryl Criaris-Bontales.*

year-old New Springville farmer still had thirty-four acres of land on Travis Avenue in 1969. He did not believe Con Ed was causing the destruction. He believed the poisonous air was coming from the New Jersey industries that lined the Arthur Kill River. Bifulco often gazed at the rows of recently built houses that bordered his property. He knew his days as a Staten Island farmer were coming to an end. His situation was exacerbated by the lack of hired help. Island farmers often hired migrant laborers. Many farmers complained that once a laborer saved enough money, he would often go home before the growing season was over. Bifulco's problem with air pollution was not a recent one. The poisonous air from New Jersey factories had been a growing concern for island farmers since the 1880s. Chemical-laden clouds led to the formation of the Staten Island Growers' Association in 1931. The unified group attempted to establish some form of pollution control. Air pollution was such a problem in the early 1900s that many farmers switched to orchard produce. It was easier to grow peaches, apples, pears, cherries, and plums as the air became more toxic from pollution over time.

The Dinger farm on Richmond Hill Road in New Springville supplied the Dinger nursery. The family also operated a roadside stand on Richmond Avenue on land that is now part of the Staten Island Mall. The farm was owned by brothers August and Frank Dinger, who were also businessmen and contractors. The siblings operated the popular Shoals Restaurant and Marina and the Shoals Fishing Station in Great Kills. The farm was started around 1922.

Henry W.D. Mohlenhoff, who was of German descent, began the Mohlenhoff family farming tradition on Staten Island in 1911. They were known for having created a special strain of celery that was well-liked by the public. The first Mohlenhoff greenhouses went up in the 1930s in response to the destructive air pollution floating over from New Jersey. In 1960, the family completely ceased growing vegetables outdoors as the pollution problem had only gotten worse. Instead, they expanded into retail flower and plant production in 1964. The air pollution that was curbing agricultural growth was actually supporting the evolving development of Staten Island in the 1960s. It was another reason for farmers to sell their land.

The Catholic Worker Farm at 469 Bloomingdale Road in Pleasant Plains was one of the first farms sold. Purchased in 1950, the twenty-two-acre farm had a barn, a greenhouse, chicken coops, pond, woods, and agricultural fields. Dorothy Day, a woman who is being considered for sainthood by the Catholic Church as of this writing, was interviewed at the time of the sale in 1964. She noted that the farm was inadequate to support all that was expected of the organization. It did provide important opportunities for the less fortunate to learn farming skills, but the movement required more from the site. It needed to serve as a school, a retreat house, a residence for the elderly, a shelter for expectant mothers, and a place for the study of religion, society, war, and peace. The lack of space was forcing the group to turn away those in need, so the operation was moved to Tivoli, New York. Another factor taken into consideration was the population boom that was occurring on the island. Eventually, the isolated farm would be inundated with modern-day noise and activity. The site had been chosen as an escape from such nuisances. Produce from the farm went to feed the needy at the Catholic Worker house of hospitality in the Bowery. Now, a developer planned to build houses on the acreage. The Catholic Worker Movement bought the farm in 1950 for $16,000. The developer bought it for $170,000 in 1964.

Pollution, soil depletion, labor costs, and taxes were making farming less profitable. For these discouraging reasons, selling agricultural land became more attractive as the decades progressed. Those seeking to profit from the purchase of open space were further thrilled that large estates were becoming less practical for the affluent, so some of their acreage was becoming available for subdivision. Developers were enthusiastic about the future. By 1959, there were twenty farms in the borough. At this time, a buyer needed $2,200 to purchase one acre of Travis farmland. Five years later, in 1964, that same acre was selling for $19,000. Money in the bank was better to most people than bending over under a broiling sun picking pole beans and tomatoes. In 1962, twelve farms were operating.

A 1964 three-part series in the *Staten Island Advance* focused on the island's disappearing farms. In the final article, Phillip Brittain wrote,

> *So far, the change has been noticeable only because the neat rows of vegetables and the men on tractors are missing. The land is in a state where its value zooms just for standing still. It is assumed that in the not-too-distant future houses will spring up in rows as straight as the celery used to be. A shopping center will take form here and there and acres of black pavement will cover the ground.*[1]

Chapter 2
PRE-BRIDGE INDUSTRY ON STATEN ISLAND

The Industrial Revolution officially arrived on the island around 1818, when a dyehouse and printing works was established in West New Brighton. It eventually became known as the New York Dyeing and Printing Establishment, and its success spurred the growth of other manufactories on the island, especially along the North and East Shores and, to a lesser extent, the west shore.

Among the companies that flourished were the Bechtel Brewery (1853), B. Kreischer Brickworks (1854), S.S. White (originally founded by the Johnston Brothers circa 1865), Rubsam and Horrmann's Atlantic Brewery (circa 1870), the American Linoleum Manufacturing Company (1874), Atlantic Terra Cotta (1898), Procter and Gamble (1905), and many other concerns. New Brighton saw the successful establishment of J.B. King and Company, which processed raw gypsum rock, in 1876. United States Gypsum bought the business in 1924. It evolved into a significant enterprise. Shipbuilding and repair were expanding industries on Staten Island at the turn of the nineteenth century. They underwent terrific growth, especially as World War I unfolded. In 1938, Bethlehem Steel purchased a former marine repair yard in Mariners Harbor. Eventually, 1,200 individuals toiled at the site as it grew in response to World War II.

Fireworks, pigments, soda water, wallpaper, biscuits, confections, and more were manufactured on the island during the nineteenth and twentieth centuries. Many of these products were known beyond the borough. Some held national reputations.

S.S. White Dental Manufacturing, Prince's Bay (undated). *New York Public Library.*

As World War II progressed, many manufacturers experienced rapid growth and vast profits as they switched to military products. When the war ended in 1945, there were 166 factories on the island. Approximately 6,080 people were employed annually.

In 1952, 12,238 individuals were employed at 252 factories on the island. But as the decade progressed, some of these companies began to fail, while several of the larger businesses moved to New Jersey. Taxes were lower, and modern production facilities could be built and then expanded on available open space, while transportation infrastructure facilitated the quicker movement of products.

Of the 142 manufacturing businesses operating in 1964, Procter & Gamble had one of the largest payrolls, with 1,370 names. Eventually, the P&G factory site would be called Port Ivory. Soap flakes, bars of Ivory and Lenox, shortening, juice, cake mix, and more were produced during the company's lengthy stay in the borough. Procter & Gamble ceased production there in 1991.

Surviving island concerns saw the Verrazzano-Narrows Bridge as a way to improve business and as an enticement for skilled workers. In the long run, this hope was not realized. Actually, the blue-collar manufacturing jobs that had been abundant on the island and throughout the city and that were usually well-paying union jobs began to disappear. These jobs had once provided salaries that allowed many single-income families to purchase a home in the borough.

While various factories were failing in the 1960s, some believed that Staten Island was the perfect locale for future industry. On March 27, 1967, the commissioner of the Department of Commerce and Industry spoke at the Staaten restaurant. He claimed that industry was leaving the city because of a lack of space for expansion. He believed Staten Island's open space was the answer. The commissioner was heralding an innovative plan proposed by Mayor John Lindsay, who was calling for an Economic Development Administration.

Chapter 3

THE FIRST REAL ESTATE FRENZY

Tradition states that George William Curtis—famed orator, author, editor, and Staten Island resident—said, "God might have made a more beautiful place than Staten Island, but He didn't." Real estate brokers took up the phrase during the late 1800s through the 1920s. (It has never been verified that Curtis wrote or even uttered these words.)

Realtors, like farmers, knew that the earth beneath their feet was truly the island's greatest resource. Staten Island is not large. Shaped like a diamond, the area encompasses 60 square miles. It is 13.9 miles long and 7.3 miles wide. It is and always has been the least populated borough of New York City.

In 1898, residents voted to be included in the city. As a result, a number of developments were seen in transportation, utilities, and public services in the borough. A major improvement in ferry service occurred in 1905, when the city took over the ferry run between Saint George and Whitehall in Manhattan. Five recently built boats were also provided by municipal benefactors that year. This was the first and only ferry service operated by the city. In this case, the voice of Staten Islanders had been clearly heard when an uproar took place about the ferry service provided by the Baltimore & Ohio Railroad in the early 1900s. The uproar reached a head when the Staten Island ferry *Northfield* was rammed from behind by a B&O Railroad boat at Whitehall on June 14, 1901. The resulting loss of life infuriated island residents. They were also upset because the B&O paid no attention to complaints about ferry breakdowns, filthy conditions aboard ferries, late ferries, and more.

Top: South Ferry Terminal where the Staten Island Ferry docks in Manhattan, 1896. *Library of Congress.*

Bottom: The Borough-class Staten Island ferryboat *Queens* approaching the slip, circa 1907. *New York Public Library.*

The Narrows, New York Bay, Staten Island. Currier and Ives, circa 1850s. *New York Public Library*.

The population of Staten Island hoped that consolidation with the other boroughs would improve life on the island. More than anything, residents yearned for fair treatment and equality with the other boroughs. This would be a factor in accepting the bridge proposal from Robert Moses in a few decades.

As part of Staten Island's inclusion in the city, the island's first borough president, George Cromwell, was elected. A surge in building took place at Saint George with the construction of Borough Hall. The Richmond County Supreme Court, the Saint George Public Library, the Richmond County Family Court, and eventually the 120th Police Precinct were constructed. Saint George was now the municipal hub. It replaced centrally located Richmond Town. The island's focal point would now look squarely across the harbor to the all-powerful borough of Manhattan. Consolidation brought modern fire and police departments and, in some cases, improved schools. Road construction increased, and macadamized roads became a much-heralded advertising point for realtors.

The *New York Times* and other newspapers were already highlighting the island. In an article called "Beautiful Staten Island Villages" published on April 21, 1895, the *Times* noted that more than $300,000 was spent for a borough road system. It was implied that a similar amount would be spent on pure water, gas, electric lights, and sewers. The island had both city and country benefits, according to the article, including an easy commute on the ferry, suburban streets, many churches, and schools. Another 1895 selling point: All houses were detached and sat on grassy plots, each with a garden, while neither the front nor the backyard was bricked or paved. Instead, "everywhere is the green velvet covering that nature provides." The article further stated that a flourishing social life, less crime, and the best shopping available could all be experienced. The article closed with several paragraphs on the temperance community of Prohibition Park. Lot availability, future community enhancements, self-improvement lectures, and concerts at the auditorium were all discussed to entice the public to move to Prohibition Park.

During the post-consolidation period, realtors designated Staten Island "South New York"—not that this was untrue. Ward's Point at Tottenville is the southernmost point of both New York State and New York City. Realtors were attempting to showcase the connection between Manhattan and Staten Island. The *Sun* of July 1, 1906, boldly stated that "South NY will be seen rising resplendent….90,000 people went last Sunday to South New York, formerly SI."[2] Houses and lots at Castleton Corners, Concord, Garretson

(Dongan Hills), Mariner's Harbor, Todt Hill, and Tottenville were all publicized. Builders and realtors were relentless. A rise in home construction resulted in the expansion of Sunnyside, New Dorp, and parts of Tottenville.

Improvements made in the borough and promises of future improvements led to an increase in population between 1900 and 1927, when the number of residents went from 67,201 to 145,723. This occurred primarily along the North and East Shores and in the well-established neighborhoods of Clifton, Stapleton, Tompkinsville, Saint George, New Brighton, West New Brighton, Port Richmond, and Mariner's Harbor. These neighborhoods were dense with residences; workers walked to nearby businesses or industrial enterprises for their employment. Many of the houses accommodated multiple families. Main streets existed, and it was there that the populace purchased food and dry goods. Movie theaters, saloons, barber and butcher shops, small grocery stores, and other businesses were generally assembled around principal thoroughfares like Richmond Terrace, Bay Street, Canal Street, Richmond Avenue in Port Richmond, Broadway, Main Street in Tottenville, New Dorp Lane, and elsewhere. For the outlying population, a trolley car, horse, or mule—or, if you were fortunate enough to own one, a car—was all one needed to procure supplies that were unavailable at the nearest crossroads.

The Staten Island Rapid Transit Railroad ran from Saint George to Tottenville in the early 1900s. A spur at Clifton existed to deposit travelers at the seaside location of South Beach. On the North Shore, a passenger train and a freight railroad ran from Saint George to a rail bridge at Old Place. Here, goods crossed the Arthur Kill River to New Jersey for further conveyance.

The years 1907, 1908, 1921, and 1924 all saw proposals for the installation of tunnels to run between Staten Island and New Jersey, Manhattan, or Brooklyn. These proposals led to increasing interest in Staten Island real estate. One of the proposals was based on a tunnel with two tubes. One was referred to as the south tunnel; the other was the north tunnel. As of late August 1922, borings for a connection between 86th Street in Brooklyn and Tompkinsville had begun. This was the northerly route. The southerly route would run from Rosebank to 100th Street in Fort Hamilton.

Movement on the subway tunnel set off a frenzy of advertisements for new homes on the island. Pages of ads for lots, houses, and apartments were featured. One bold announcement proclaimed: "Subway to Staten Island. Come out and buy now—before price advance! Houses—plots—lots and acreage."[3] Another ad declared that residences were available for purchase and could be lived in during summer or winter. No doubt this

Area of business around Staten Island Savings Bank at Beach and Water Streets, Stapleton, pre-1923. *New York Public Library.*

Four early twentieth century plans to install a subway across the Narrows never came to fruition. *Bay of New York*, circa 1820s. *New York Public Library.*

was noted because some Manhattan residents viewed Staten Island as a summer getaway. Midland and South Beaches both had flourishing cottage communities. Advertisers insisted that Grant City, New Dorp, and Great Kills "may properly be named the RIVIERA OF NEW YORK CITY."[4] The exaggerations continued with the statement that these waterfront locations were protected from the cold winds of winter by a range of hills. Of course, Emerson Hill, Todt Hill, and Lighthouse Hill are miles away from the beach, thus affording little shelter from the weather.

The tunnel or "subway boom" led to the purchase of thousands of Staten Island acres. One realtor stated that the subway boom was part of Mayor Hylan's program to provide a freight and passenger tunnel that, presumably, would have brought business to Staten Island's modern East Shore piers. This led to a real estate rush as well, but Governor Al Smith and Mayor Hylan had differing opinions. In 1925, the Nichols-Hofstadter Bill was passed by the New York State Assembly and signed by Governor Smith. It ended the "subway boom" as there was a clause that disallowed use of the same underground tunnel for freight and passenger traffic.

Realtors and developers gushed over the fact that Hylan Boulevard was over eleven miles long in 1928. They stressed that 80 percent of Staten Island's 405 miles of roads were paved at that time. These facts, along with the soon-to-be-opened Goethals Bridge and the Outerbridge Crossing to New Jersey, plus the anticipated construction of the bridge to Bayonne, were making Staten Island the most promising of the five boroughs. Realtors were "priming themselves for a busy season" and an "unprecedented" year.[5] According to one island realtor, from 1922 to 1927, Staten Island realty values jumped 110.3 percent. An increase in telephone service and public utilities and the erection of new public structures also increased property values. Politicians, businesspeople, and realtors believed that the population would reach one million in just a few years.

The Goethals Bridge and the Outerbridge Crossing were built by what is now the Port Authority of New York and New Jersey. The two spans opened simultaneously on June 20, 1928, while the Bayonne Bridge was dedicated on November 14, 1931.

Indeed, Staten Island's population did grow in response to the improvements. The arrival of immigrants also bolstered the number of island residents. The end of the nineteenth century and the early twentieth century saw large numbers of Italians and Greeks settling on the island. The possibility of establishing a farm was a great enticement to both groups. The shipbuilding industry was still luring Scandinavians to Staten Island's North

Greek immigrants Emmanuel Kriaris and his cousin Panayiotis Chletsos, pictured circa 1910, experienced success in the United States. *Courtesy of Dr. Cheryl Criaris-Bontales.*

Shore, while Hungarians were able to work in the brick factories on the west shore. Poles sought work in Elm Park and Travis, while Eastern European Jews settled in the retail areas of the borough. Stapleton, Tompkinsville, Manor Heights, and Sandy Ground saw increases in the population of African Americans, many of whom had connections to the rural South.

But things changed radically when the Great Depression hit. Beginning in 1929 and throughout the 1930s, poverty and upset spread. Property owners could not pay their real estate taxes, so thousands of Staten Island plots realized foreclosure. With that, New York City became the largest landowner on Staten Island. As anyone who has walked through wooded areas on the South Shore knows, sidewalks could be found there up through the 1990s at various locations where houses and small developments were planned.

When the Depression ended and World War II was finally over in 1945, the City of New York wanted increased revenue from the outlying boroughs. Plans to increase the population were developed. This was a job for someone who knew how to get things done. When the Triborough Bridge Authority and the New York City Tunnel Authority merged in 1946 to form the Triborough Bridge and Tunnel Authority, one man was poised to see that Staten Island could literally be found on the map. This man was Robert Moses, the chief of the new Triborough Bridge and Tunnel Authority: a man who preferred bridges—and despised subways and tunnels; the man who saw to it that the Triborough, Throgs Neck, Marine, Henry Hudson, Cross Bay, and Bronx-Whitestone Bridges would be built; the man who would go on to erect the Verrazzano-Narrows Bridge between Staten Island and Brooklyn.

It should be noted that in 1928, a proposal to build a bridge from Staten Island to Brooklyn was submitted to the United States Department of War. This proposal came from the Interboro Bridge Company, a group consisting of businessmen from both boroughs. Tolls would support the $60 million suspension bridge. Interboro suggested that financial backing for the bridge had been secured, but Interboro would not reveal the identity of the backers. The span would handle only vehicular traffic; there would be no trains. It was later revealed that building the bridge would not have been possible since the technology needed to produce sufficient steel cables to support the bridge had not yet been developed.[6]

Another proposal for a bridge across the Narrows in 1929 called for bridge towers with observational balconies where the public could take in views of the New York Harbor. In addition, "a carillon of bells [would] peal out anthems of liberty to those who enter through this nation's gateway; and

Robert Moses with a model of the Brooklyn Battery Bridge, 1939. *Library of Congress.*

a beacon of light [would] send out its beams as a symbol of liberty to guide, welcome and inspire those who have crossed the seas to come to our shore."[7] This proposal also lacked a subway line.

Three decades later, the idea of installing a rapid transit line on the "Narrows Bridge" was still ignored. The *New York Times* asked why it was not possible to design the bridge so tracks could be installed at a later date. Robert Moses responded in a 1956 letter to the editor. He believed that "rapid transit facilities on the bridge would add substantially to its cost." Moses further believed that "rapid transit to Staten Island is not required now or in the foreseeable future" and that since the rapid transit of the time was already in the red, the city could not afford to build a system on the Narrows bridge. In addition, Moses stated that the train lines in Brooklyn could not accommodate additional trains during rush hour. Most notably, Moses wrote, "Dumping capital costs of city rapid transit and private suburban train commutation on the backs of automotive owners and drivers who pay their way is no solution of urban rail transportation." Moses closed his letter with: "We can be wrong, but not in this case."[8]

Chapter 4

PLANNING FOR THE "NARROWS BRIDGE"

Those who can, build. Those who can't, criticize.
—Robert Moses, New York Times, *July 30, 1981*

It was 1924, and Robert Moses was developing plans—plans he believed would improve Long Island, New York City, and various other sections of New York State. His public works projects would eventually include highways, bridges, power plants, playgrounds, tunnels, beaches, parks, housing, civic centers, zoos, exhibition halls, and two World's Fairs. Most of his campaigns revolved around the automobile. Nowhere was he more influential, and controversial, than in building roadways for automobiles. Some declared he did not build; he rammed roadways through the landscape—mainly the landscape of New York City. Moses had ample power. At one time, Robert Moses simultaneously held twelve appointments. (He was never elected to any position.) His many titles included New York City parks commissioner, chairman of the New York State Power Authority, chairman of the New York State Parks Council, head of the New York State and Long Island Parks Commissions, commissioner of the New York City Planning Commission, among numerous others. One of these appointments alone gave Moses more power than the average politician could ever realize. His influence was most felt as chairman of the Triborough Bridge and Tunnel Authority (TBTA) from 1934 to 1981.

As Robert Moses's career wound around to a crossing at the Narrows, certain members of the public and certain politicians began questioning him

The New York Pavilion at the World's Fair in Flushing, New York, on April 14, 1964. Robert Moses was president of the New York World's Fair 1964 Corporation. *Library of Congress.*

and the projects he developed, especially after the construction of the Cross-Bronx Expressway in 1950. This roadway brought about the demolition of more than one thousand apartments and the displacement of thousands of residents. In addition, a thriving one-mile stretch of East Tremont in the South Bronx was destroyed.

In 1946, the Triborough Bridge and Tunnel Authority was authorized to build a bridge between Brooklyn and Staten Island, a bridge that would not interfere with navigation or the harbor traffic that swept into and out of the Port of New York. Both Fort Wadsworth on Staten Island and Fort Hamilton in Brooklyn would be used for bridge anchorage and approach purposes. By using these federal properties, the project would not impact residential neighborhoods—supposedly. The public later realized that approach roads and highways leading to the bridge in both Brooklyn and Staten Island would impact neighborhoods.

By the 1950s, the bridge was becoming a reality. Thus, developers began eyeing the open spaces of the least populated borough. At this time, residents felt good. There were jobs, money, and a future that many—not all, but many—were looking forward to. Also looking forward to the upcoming years were realtors, politicians, businessmen, and developers. Indeed, developers would realize a prosperity unknown to most. By 1959, land speculation on the South Shore and in the Mid-Island areas of Staten Island was occurring at a phenomenal speed in anticipation of the new crossing. The purchase of vast amounts of acreage indicated that a building boom was about to

Fort Lafayette (*foreground*) and Fort Hamilton (*background, center*) where the Brooklyn entrance to the Verrazzano-Narrows Bridge would be located, 1852. *New York Public Library.*

explode. Much of this property was now in the hands of off-island developers who had purchased land at in rem sales or auctions held by the City of New York—the largest landholder of Staten Island property.

Simply put, in rem property was and is property whose owner has not paid property taxes, as occurred during the Depression. It was also land that was deemed a hazard with no sign of the situation being remedied. When the city resold these properties, a substantial profit was realized.

Much of the in rem land first purchased was repriced by off-island speculators and was now considered overpriced by Staten Island builders and brokers. According to one realtor, selling most of the land to developers and speculators, who were the city's prized customers, was "impeding the orderly development of the borough." Planners and professional men of the island believed the "in rem sales [were] conducive to wholesale construction methods, a sameness of architectural design and little or no site planning."[9] This was, in fact, one of the complaints that would come from native islanders and later residents.

One *Advance* article and an ad in the classifieds mirrored dozens of others. Both were published on June 13, 1965. The article was titled "City to Sell Land on Thursday." Sixty properties were being offered for sale at the Statler Hilton in Manhattan. Most of these sites were located on the South Shore of Staten Island. A description of one Annadale property read: "Land between Rathbun and Rensselaer Aves. measuring approximately 200 by 485 feet is being offered at $71,600." Also, "a 461-FOOT deep parcel within the 200 feet between Rensselaer and Sheldon Aves. has a minimum price tag of $67,800." Thirty-one Annadale parcels were also for sale. Other properties to be auctioned were located at West New Brighton, Fort Wadsworth, Willowbrook, Bay Terrace, Huguenot, and Princes Bay.[10]

By 1964, the City of New York owned 60 percent of Staten Island. The city made more than $37 million auctioning off parcels between 1960 and 1964. That is roughly $397 million in today's dollars.

Not only were developers pleased, but so were businesspeople and many politicians since they too favored growth on Staten Island. The Staten Island Chamber of Commerce believed the bridge would immensely benefit the island. A contemporary logo featuring the bridge and the words "Bridging the Future" was planned for the chamber's newsletter. One board member urged forty-five pharmaceutical companies to open plants on the island. He believed the borough could serve as a distribution point for customers between Washington, D.C., and Boston. Understandably, this was the mindset at the time.

Above: Bucolic Huguenot, site of Arbutus Lake, became a popular location for new construction beginning in the 1960s. *Undated postcard, New York Public Library.*

Opposite: The Staten Island tower of the Verrazzano-Narrows Bridge from Battery Weed at Fort Wadsworth, 1980s. *Photo by E.K. Criaris. Courtesy of Dr. Cheryl Criaris-Bontales.*

Signs of borough advancement were everywhere. Another 209 acres was approved for inclusion in Great Kills Park in the early 1960s. This was good news as elsewhere, ranches and Cape Cods were replacing meadows, forests, and wetlands. Sweet gum and sassafras trees alike were being decimated with little regard as to whether they were common or rare. Concern about the destruction of flora and fauna would come long after the fact. Roads were laid while septic tanks were ineptly installed on postage-stamp-sized properties that held houses fewer than ten yards apart. Where would the effluence from the tanks go? Nobody cared—that is, until the new buyers moved in. What about schoolrooms for arriving kids? Many people believed that the overpopulation of existing school would be solved by creating junior high schools. Unfortunately, only one was completed by 1960. On the border of Graniteville and Port Richmond, Junior High School 51 alone was not going to solve the problems of kids needing classrooms.

Most of the newly arrived homebuyers were from Brooklyn. Breadwinners were used to taking speedy subways to their jobs in Brooklyn or Manhattan. They were shocked. They had been told by unscrupulous realtors and developers that their new commute would take no time. Sad to say, traffic

on the bridge and the Brooklyn-Queens Expressway would never get better. In fact, it would only get worse.

Real estate prices for land and residences on the island had historically been less expensive than in the other boroughs. Around 1962, a two-bedroom unsewered ranch in Bulls Head on a plot less than a quarter of an acre sold for approximately $16,000. (Because it was on an oversized lot, the same site sold for approximately $375,000 around 1988. The house was demolished, and at least two structures with sixteen residences took its place.)

The subject of sewers was a sore one in the borough. Staten Island was not suited for septic tanks. Clay was found under most of the island. In 1960, one home builder observed, "An acre worth $20,000 without sewers would be worth 'twice as much' if it could be hooked up with sewers." He mentioned that the city was slow to approve sewer projects. A member of the local chamber of commerce said the "main problem for developers…is lack of sewers, which depends in part on the mapping of streets, which in turn depends on drainage mapping."[11]

By the mid-1950s, the Staten Island land grab had begun, and it quickly accelerated before the first spade hit to break ground for the new bridge. Many of these land buyers either represented or headed development companies that were not located on the island. Some were situated in Manhattan. Some of the investors were from Switzerland, Germany, and Italy.[12] As land was purchased, those in charge of the city coffers realized that idle city-owned in rem land on the island could bring in millions of needed dollars for the city to pay its bills, salaries, and more. As such, more and more property was put up for sale. This process went on for years.

The value of Staten Island land soon tripled. A twenty-by-one-hundred-foot Eltingville lot that was valued at approximately $1,000 in 1963, was worth $3,000 by May 15, 1965. Newly constructed high ranches or ranch-style homes could be bought for $23,000 or more at this time. This style of home would soon cost $28,000 or more. Interestingly, many of the houses built in Eltingville lacked sewer connections. For this reason, it was referred to Septic Tank Land.

A "Staten Island Home Buyers Guide" was published around 1961. The booklet begins with the quote "God might have made a more beautiful spot than Staten Island, but He never did." (The wording differs from previous versions.) The booklet is filled with advertisements to entice the house-buying public. Numerous builders advertised in the booklet. All stressed the superiority of their products and of living in the borough, using catchy phrases like "Wholesome and pleasant suburban life that Staten Island so aptly provides," "More living, comfort and convenience," "Unsurpassed location," "SI is only 40 minutes from the hustle of New York," and "Swimming Club just a short walk away."[13]

Many of the houses came with porches, air-conditioning, and full basements. One ad mentions a "science kitchen." The inside back cover features lending institutions that could assist with mortgage loans, including Colonial Federal Savings, Edgewater Savings, Tompkinsville Federal Savings, Westerleigh Savings, and North Shore Federal Savings. Richmond County Savings Bank and Staten Island Savings Bank also provided mortgages for many newly arriving island residents over the years.

Chapter 5

BUILDING THE BRIDGE

Fortune seekers were ready for the Verrazzano-Narrows Bridge—as some called it, the New Colossus, or as Robert Moses called it, the Big Bridge—to be built. Many referred to it simply as the Bridge. They still do.

The Port of New York Authority "financed the original borings at the tower sites" and the "survey to determine the actual dimensions" of the bridge. It was agreed early on that the bridge would be a joint venture of the Port Authority and the Triborough Bridge and Tunnel Authority. This changed on December 12, 1959, when the Port Authority and Robert Moses, head of the Triborough Bridge and Tunnel Authority, proclaimed that the Port Authority would "step out of the picture' as financial backer" of the span. This agreement would become official when Triborough repaid monies previously spent by the Port Authority. According to Triborough, it was "now able to meet the cost of financing the project." When asked who called for the change, neither agency would respond. When asked why the change was occurring, neither agency would respond.[14]

Ground was broken for the bridge on August 13, 1959, at Fort Wadsworth. The attendees numbered three thousand. It was the most important event in Staten Island's history. Robert Moses smiled as he and others tossed the first chrome-covered shovels of dirt away. The steamy summer day featured an unannounced plane trailing overhead with a sign that read "Name it the Staten Island." In honor of the ground-breaking, a twelve-foot ice carving of the bridge was on view. By the close of the event, the carving had succumbed to the heat, and it was little more than two tiny lumps of ice.[15]

Historic Fort Wadsworth (as seen in this circa 1910 postcard) served as the ground-breaking site for the Verrazzano-Narrows Bridge on August 13, 1959. *New York Public Library.*

There wasn't a car-owning family on Staten Island that did not visit the bridge during its construction. A camera was usually present, too, to document the bridge's advancement. Some believed that numerous international visitors arrived in the United States solely to see the bridge being constructed. All who came to see the construction were referred to as Bridge Worshippers, and their destination in the early 1960s was the Narrows. It was said that hundreds visited the Brooklyn side, while thousands visited the Staten Island side at Von Briesen Park or Fort Wadsworth. One overcast winter Saturday in 1964 saw eighty-eight cars at Fort Wadsworth to view the expanding span. The next day, a sunny Sunday, there were an astounding 651 cars at the same location. By late April 1964, more than 1,500 individuals were visiting Fort Wadsworth on weekends to see the progress of the bridge.

Families were not the only visitors. Organized Scout troops, civic organizations, school classes, and photography classes were showing up to view the spectacle. Some took guided tours with Fort personnel that focused on the bridge. As the great monstrosity advanced, family photos with the bridge in its various stages of completion served as background scenery. Photographing Mr. Moses's creation was so common that the *Advance* of January 22, 1964, announced a photo contest featuring the

Top: Frank J. Salmon at Fort Wadsworth during the construction of the Verrazzano-Narrows Bridge, 1963. *Photo by James J. Salmon. Author's collection.*

Bottom: James P. Salmon at Fort Wadsworth during the construction of the Verrazzano-Narrows Bridge, 1963. *Photo by James J. Salmon. Author's collection.*

bridge. Amateur photographers could submit images through October 1, and the winning entries would be featured in the newspaper when the bridge opened in November. Three prizes would be given to the winners: a $100 U.S. Savings Bond for first prize, a $50 bond for second prize, and a $25 bond for third prize. There would be three honorable mentions. First prize went to Steven Wasserman of Graniteville, a motion picture major at City College. His photo showed a carefree fisherman, complete with pole and high rubber boots, seated in a chair with the majestic bridge as a backdrop.

As for building the bridge, the men who worked on it referred to themselves as Iron Workers, although the material used was steel. The bridge "employed a daily average of 1,200 men for five years." All told, "10,000 men worked on the bridge at one time or another."[16] Some lived in Brooklyn; others traveled across the United States from one bridge job to another. Native Americans from the Caughnawaga Reservation near Montreal, Canada, who commuted home on weekends, labored on the project.

Three men died during construction. Some believed that had under netting been in place, their deaths might have been averted. The first victim was Paul Bassett of Tenth Street, Brooklyn. On August 24, 1962, the experienced ironworker was working 400 feet above the Narrows on the Brooklyn tower. He was about to bolt rivets when his ladder slipped and he fell 150 feet through an elevator shaft. Bassett was pronounced dead at Coney Island Hospital.[17] The second death occurred on July 13, 1963, when Irving Rubin fell from an approach road. He succumbed to his injuries.[18]

Gerald McKee was the last man killed on the bridge. He was nineteen years old when he fatally slipped on a catwalk and fell three hundred feet to his death on October 10, 1963. The young man landed in the water. McKee's fellow ironworkers stopped their tasks to grieve the loss. The stoppage signaled a traditional mourning period. It also allowed for an investigation into McKee's death. Gerald was an apprentice ironworker. His two brothers, also ironworkers, were working on the bridge at the time. James McKee, age twenty-three, was also an apprentice, while his brother John, age twenty-two, was a journeyman. The brothers made their way to the Brooklyn side as Gerald's body was being retrieved from the water. A coworker said of Gerald, "He was as careful as the rest of us."[19]

Bridge workers walked off the job over safety concerns on December 2, 1963. They wanted safety nets. American Bridge, a division of the United States Steel Corporation, said that "nets were unnecessary and that safety ropes and belts were adequate protection."[20] The company also insisted that

Looking toward Brooklyn and Manhattan from the Verrazzano-Narrows Bridge, after 1968. *Photo by Jet Lowe. Library of Congress.*

installing nets was hazardous. The men worked between 230 feet and 700 feet above the Narrows. Construction recommenced on December 9, 1963. The men had "won" the precious nets.

The ironworkers who labored on the massive bridge were a family. Like Gerald McKee and his brothers, many of them worked alongside family members. James McLemore, age forty-six, and his son James, age twenty-three, lived on Latimer Avenue in Travis. Each morning, they drove to the bridge for jobs that took them sky-high over the Narrows. Here, everyone walked carefully. Everyone was said to be scared, regardless of what they said. The younger McLemore was certainly grateful for the improved protection when he fell from the suspended roadway to the "big circus-like safety net below" during March 1964. Except for a few bruises, he was unharmed. The bridgemen discussed the wind that rocked the bridge, which made the suspended steel shift three to four inches. At this time, the elder McLemore was walking with a cane. He'd had hip surgery owing to an on-the-job accident. He had hoped that his son would go to college, but after a summer spent apprenticing on a Chicago highway bridge, the younger man knew he had found a career that he would love. Mother and wife Jean didn't like to think of her husband and son laboring on the bridge, nor did she want to visit and see them working. She feared that she would be a distraction and they would slip and fall.[21]

The Verrazzano-Narrows Bridge was designed by Switzerland native Othmar Ammann, one of several New York City spans he was involved with. These included the George Washington Bridge, the Bayonne Bridge, the Bronx-Whitestone Bridge, and the Throgs Neck Bridge. Ammann is quoted as saying, "It is a crime to build an ugly bridge."[22]

The giant towers of the Verrazzano-Narrows Bridge stand 693 feet high. The entire structure is so massive that the curvature of the earth needed to be considered during its design. For this reason, the tower tops are 1⅝ inches farther apart then are their bases. There is enough wire in the cables of the bridge to reach halfway to the moon or to be wrapped five times around the equator.[23]

During the blistering hot days of a New York City summer, the cables of the bridge expand. This makes the roadway twelve feet lower in summer than in winter. After the roadway was put in, the arc of the bridge's cable descended approximately twenty-eight feet. A lower roadway was opened on June 28, 1969. This allowed for twelve lanes of traffic.[24] This addition was eleven years ahead of schedule. At the time, roughly twenty-eight million vehicles crossed the bridge per year—far above the TBTA's original projection. The final cost for building the bridge was $325 million. This included construction of the bridge approaches, reconstruction of the forts, purchasing property, relocation costs, administration, and engineering. It was paid for by selling revenue bonds to the public and repaying bondholders from the income generated by TBTA facilities. There was never any mention of the bridge becoming free to cross after its construction was paid for or the idea that it should be free since it is part of the interstate highway system; these are urban myths.

On a tiny spit of land off the Brooklyn shoreline, a small but significant fort had stood for approximately 137 years. Known as Fort Lafayette, it was demolished to make way for the Brooklyn tower of the massive bridge. Built between 1812 and 1822 as part of a harbor defense system in response to the War of 1812, the fort was used during the Civil War. In fact, Fort Lafayette imprisoned "high value" Confederates. The last prisoner of war is said to have been executed there on March 25, 1865. Little interest in the fort's antiquity was displayed when the island was tested for its ability to hold the Brooklyn tower. One journalist wrote that the "groundwork for what will be one of the wonders of modern science and what will be an even greater landmark than the picturesque, brownstone-faced military installation" was

Left: Another of Othmar Ammann's creations was the George Washington Bridge, seen here in 1937. *Photo by Berenice Abbott. New York Public Library.*

Below: The lower roadway of the Verrazzano-Narrows Bridge opened on June 28, 1969. *Photo by Jet Lowe, 1969 or later. Library of Congress.*

taking place in 1959. The fort was described as abandoned and useless, nothing but a home for pigeons "that nested in the cavernous stone-walled rooms and gun embrasures." Workers were busy extracting samples from far below the island to determine its capability; all were pleased to report that a "beautiful sample" of granite gneiss had been pulled from the rocky belly of the island.[25] Not even the legend of a tunnel between the fortification and Fort Hamilton in Brooklyn proper could save this vintage shrine. Except for an article in the *New York Times*, the magnificent fort was being removed with

Fort Lafayette off Brooklyn, 1861. The fort was removed to make way for the Brooklyn tower of the Verrazzano-Narrows Bridge. *New York Public Library.*

little fanfare. In fact, the entire island was removed. Fort Lafayette was wiped off the map. Historic stones and bricks were discarded into scows that were quietly dumped off the shoreline of Staten Island.

Fort Wadsworth was subject to disfigurement and demolition by the construction of the bridge. On August 18, 1959, a wrecking ball arrived to destroy a two-story Italianate structure that overlooked the parade grounds. Built in 1870, it was a private residence before the property became part of the military reservation in 1893. Around this time, it was to be used by post commanders.[26]

The bridge's approach from Staten Island cut a 350-foot swath through Fort Wadsworth. In 1959, it was expected that approximately six major buildings and twenty temporary structures would be demolished. It is important to note that the site first became a military outpost around 1663, when a blockhouse was built by Dutch settlers.

In general, most Staten Islanders were not opposed to the bridge. This was in contrast to a number of Brooklyn residents. Many were appalled at the number of people who would be displaced by the span, its approaches, and its connecting highways. Islanders believed the bridge would spur growth in population, business, industry, and finance and put Staten Island on equal footing with the other boroughs. Island residents have always believed theirs is the forgotten borough. There is no subway. The Fresh Kills Landfill was foisted on the island in 1948, and in the early 1960s, its closure was nowhere in sight. Staten Island was not connected to the rest of New York City physically or mentally. Only the ferries "joined" the borough with its counterparts, and these were not permanent connections—emotionally or visually. Staten Islanders always had a "lack of identification…with the rest of New York City." This was grounded in physical separation from the "mainland." It was termed "islandness."[27]

The Staten Island Chamber of Commerce organized a group to present the borough's desire for the Narrows Bridge before the City Planning Commission on June 30, 1958. Meanwhile, Bay Ridge residents were gearing up to show their opposition to the span. What most upset residents was the destruction of numerous homes along the bridge's feeder routes: Sixty-Fifth Street and Seventh Avenue. Homeowners and their allies called for a route on the bay side of Belt Parkway. But Robert Moses and his allies claimed that would be too costly and federal funding would not be provided for such a route. Moses informed all that if his Brooklyn route was not built as planned, the bridge would not be constructed.[28]

The June 30 meeting went on for an alarming nine hours. When it was finally over, members of the City Planning Commission "reserved their decision" on the Brooklyn approach to the bridge for a later date. The appeals by opponents were emotional, off topic at times, and often evocative. Some referred to the Narrows Bridge as "Moses' folly," a "menace to navigation," and a project "nobody wants."[29] Robert Moses was called a

New York Bay (From the Telegraph Station), 1838. The telegraph station was the original location of a seventeenth-century Dutch military outpost and the future site of Fort Wadsworth. *New York Public Library.*

The Fresh Kills Landfill opened in 1948. It was the bane of island residents until 2001. *Photo by William Criaris. Courtesy of Dr. Cheryl Criaris-Bontales.*

variety of descriptive names, including "city destruction coordinator" and "dictator." The meeting began at ten o'clock in the morning. Staten Island's bridge supporters were not heard until after six o'clock. A member of the Staten Island Real Estate Board defended Moses by calling him a gentleman. He claimed the City of New York was fortunate to have a man of Moses's caliber on the job. One builder believed that Moses should receive a medal for his services to New York City.

At a Board of Estimate meeting on September 27, 1958, that lasted a massive fifteen hours, a number of religious leaders from Bay Ridge spoke against the Brooklyn approach. One monsignor said building the bridge bordered on the "sacrilegious." He stated that Robert Moses was "like a little boy who threatens to take his bat and ball and go home," in reference to Moses saying the bridge would not be built if the Seventh Avenue route was changed. The monsignor continued, "He should be allowed to go home. We can get another ball and bat—and we can get another player." Another clergyman believed that Moses wanted the bridge to stand as a monument to himself. He suggested that Moses hoped the span would one day be renamed in his honor. One builder spoke. He noted that his "firm was ready to spend $50 million on the island for new homes, apartment houses and shopping centers. He invited persons in Bay Ridge who might be displaced by development of the Seventh Avenue approach to move into the island homes his company built."[30]

The people of Bay Ridge had a disappointing New Year's Eve in 1958. On that date, the Board of Estimate approved construction of the bridge. Plus, the approach from Brooklyn would ram through Seventh Avenue. The resulting highway would be named the Gowanus Expressway. The only nay vote came from the Brooklyn borough president.

Robert Moses agreed to assist displaced persons who lived within the approaches of the bridge and the Staten Island Expressway. For this reason, a relocation office was established at 603 Manor Road at Castleton Corners. Experienced realtors were stationed at the site to explain condemnation, relocating houses and families, and methods for purchasing other homes. In 1958, the Port Authority estimated one hundred houses would be moved.

One of the primary motives for building the bridge was to circumnavigate traffic away from Manhattan. It was also built as an eastern connection for the southern and northern portions of the U.S. interstate highway system. Moses promised that the bridge would allow for a trip of forty minutes between central Staten Island and Times Square. What it really allowed was access between Staten Island, New Jersey, Brooklyn, Queens, and Long

Looking north of Eighty-Sixth Street where houses were demolished to make ready for the Verrazzano-Narrows Bridge approach in Brooklyn, 1961. *Library of Congress.*

Island. As time went on, the plan for a quick and easy route disappeared like the foggy mist that often surrounds the mighty bridge.

People were attracted to the island for many reasons, including lower taxes, houses that were less expensive, less crime, less traffic, more parking, and newer neighborhoods. Most importantly, island schools had excellent reputations. This exacerbated an existing Staten Island problem: Schools were already overcrowded.

Robert Moses's mind was in constant motion. Before, during, and after the opening of the Verrazzano-Narrows Bridge, Moses was not only connecting the directions and routes of his roadways to each other and beyond but also creating the need for these roadways and bridges. In 1965, he wrote an in-depth report called *The Framework of Staten Island*. The report was showcased in a newspaper article, where it was called a "plan for the future development of roads and recreational facilities" on the island.[31] Interestingly, Moses wrote that the Triborough Bridge and Tunnel Authority and the Port of New York Authority "accepted major

A house being moved to Foote Avenue along Clove Road, Sunnyside, January 15, 1959. The Emerson Drive home sat in the expressway's path. *From the* Staten Island Advance.

responsibility for the impending growth of Staten Island" beginning in 1934. It is now 1965, the bridge is open, and Moses declares that he will "continue to help coordinate plans for the framework of transportation and incidental improvements necessitated by the opening of the Verrazzano-Narrows and its approaches and connections." There is hardly any mention of public transportation other than to note that the island had no local passenger train service except that which exists, and still exists, between Saint George and Tottenville. The Triborough Bridge and Tunnel Authority chairman concedes that the Staten Island Ferry is not yet obsolete; instead, he states that the ferry "boasts the cheapest and finest five cent ride anywhere, an anachronism of anachronisms defying all logic and economics, the untouchable march of nickels."[32] He is referring to the world-famous nickel ferry fare that was in place for decades. In Moses's mind, Staten Island was at a "crossroads." It could "no longer live to itself." It could no longer be isolated. Moses plainly states that the "Big Bridge" will remove "rural remoteness" and provide a "unique opportunity" to preserve the country in the city.[33] There is no mention of what borough residents desired.

Naming the bridge led to numerous arguments. It was originally called the Narrows Bridge. John LaCorte of the Italian Historical Society of America had a lengthy discussion with Robert Moses about another suitable name. An agreement was reached that the Italian explorer Giovanni da Verrazzano would be honored. It is believed that he was the first European to see the land mass that would become Staten Island in 1524.

Much has been made of the correct spelling of the explorer's name. Two *z*'s or one *z*? Some claim Moses's secretary left out a *z* in the paperwork. Some claim Moses had one *z* removed because "Verrazzano-Narrows Bridge" was too long. Some claim it does not matter as Verrazzano was historically spelled with both one *z* and two *z*'s. In any case, Governor Andrew Cuomo officially changed it to include two *z*'s in 2018.

Island representatives had asked for resident discounts early on, but they were not forthcoming. The TBTA did promise "commutation books," but they were for convenience, not for saving money. The booklets would allow for ten trips to Brooklyn at a cost of five dollars—end of story. Discussions regarding tolls would go on for years. They continue today.

Chapter 6
THE OPENING OF THE "NEW COLOSSUS"

To celebrate the bridge, the U.S. Postal Service unveiled a five-cent stamp on August 22, 1964, that featured an image of the great span. Some said it honored the "Eighth Wonder of the World." The first-day issue of the stamp was set for opening day of the bridge on November 21, 1964. The Staten Island Philatelic Society was also taking orders for a special Verrazzano-Narrows Bridge first-day cachet. These envelopes survive in the collections of many who were alive when the Verrazzano-Narrows Bridge opened.

When opening day finally arrived, Robert Moses called the bridge a "triumph of simplicity and restraint."[34] It was insured for $160 million. The passenger toll for automobiles was set at $0.50 one way. Automatic toll-taking machines were installed at six lanes. Two center lanes could be reversed depending on the flow of traffic.

Interesting comments were made by Robert Moses regarding an episode of *Eye on New York* that focused on the changes island residents would experience owing to the bridge. Moses stated, Staten Island Borough President Albert "Maniscalco and his friends on Staten Island want to think the bridge provides a connection between the Island and Brooklyn and a faster route for Islanders. They are mistaken. It is one thing, the last link in the vast arterial system of New York."[35] Moses lashed out at the episode, claiming it depicted the builder and the city as not having prepared a plan of development for the borough. Moses took the episode as a personal attack. He stated that it showed "nothing but scandalous officials and citizens neglect,

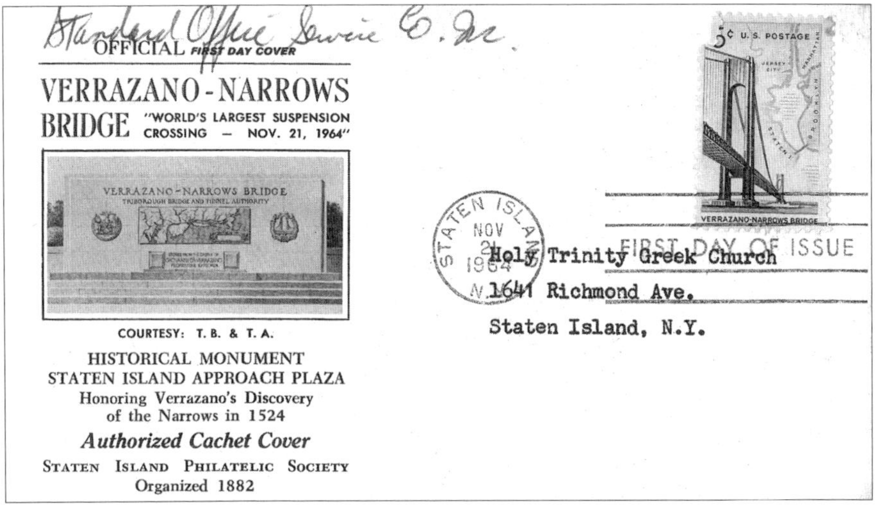

First day cover, a collectible keepsake issued by the U.S. Postal Service, commemorating the opening of the Verrazzano-Narrows Bridge, November 21, 1964. *Courtesy of Dr. Cheryl Criaris-Bontales.*

vandalism, speculation, and in the end boom and bust."[36] He lambasted the producer for not featuring accomplishments that had been achieved due to the bridge, including parks, beaches, Richmondtown, parkways, railroad grade eliminations, reclamation projects, and higher zoning standards. Per Moses, the presentation pitied the passing truck farms, towering trees, and the absence of drainage.[37]

Perhaps to placate its residents, Fort Hamilton in Brooklyn was chosen for the opening ceremony of the Verrazzano-Narrows Bridge. Moses informed all that 1,500 people would participate. This number was supposedly another reason why Brooklyn was chosen. Most had expected the ceremony to take place on Staten Island, where the toll booths, administration building, and control building were located. The Staten Island Chamber of Commerce and borough residents were very disappointed. At the urging of Borough President Maniscalco, the chamber, civic groups, and a host of residents protested. Moses surprised all by switching the ceremonies to the Staten Island side. This was one of the few times that Robert Moses was persuaded to change a plan.

The *Staten Island Advance* believed that the bridge would lead to a renaissance of the island. A modern era would be realized that featured growth, expansion, prosperity, and rebirth. The *Advance* explained that the bridge meant many things to many people. "To the Triborough Bridge and

Tunnel Authority it means adding another link to its wondrous facilities that serve the world's largest and busiest city."[38]

The Verrazzano-Narrows Bridge opened on Saturday, November 21, 1964. It was a frigid day. The festivities commenced with a parade from Hylan Boulevard near the Staten Island Expressway to a point near the toll plaza of the new bridge. Island churches heralded the span by sounding their bells at eleven o'clock, while cannons at Fort Wadsworth and Fort Hamilton blasted in celebration at twelve thirty. The official program began with a ribbon cutting on the Brooklyn side. Mayor Robert Wagner, Staten Island Borough President Albert Maniscalco, Governor Nelson Rockefeller, Brooklyn Borough President Abe Stark, and Governor Richard J. Hughes of New Jersey were on hand, as was Italian Ambassador Sergio Fenoaltea. Of course, Robert Moses was present. In fact, he was the master of ceremonies.

After the ribbon was severed, Robert Moses, federal, state, and local officials proceeded by automobile to the Staten Island toll plaza for music,

Left to right: Albert Maniscalco (*at the podium*), Robert Moses (*with hat, leaning forward*), Nelson Rockefeller, and Cardinal Spellman chatting at the opening ceremony of the Verrazzano-Narrows Bridge, November 21, 1964. *From the* Staten Island Advance.

speeches, and congratulatory backslapping. Afterward, the motorcade crossed back to Brooklyn for the party.

An army artillery salute, fireboat water displays from four navy destroyers, a marine parade of coast guard vessels, New York City Police Department boats, and commercial towboats all applauded the opening of the great bridge that day.

A three o'clock in the afternoon, the first toll for the span was collected. Six days before, at noon on November 15, 1964, seven young men pulled their fashionable Cadillac as close to the majestic crossing as they could get without being chased away by authorities. They wanted to go down in history as the first to cross the bridge, and they were successful. Ronald Sacoff, Frank Picone, Richard Ramaglia, Anthony Lenza, Benjamin Goldsmith, George Scarpelli, and Robert Caplan would take turns eating, sleeping, and passing time until the roadway officially opened. They wiled away their wait singing songs while Ramaglia played his guitar. After they changed into tuxedos, theirs was the first car to cross the span. The classy Cadillac was followed by fifty-two other cars. A marching band performed while an ocean liner let off a magnificent blast. That same day, the Staten Island Expressway and the Willowbrook Expressway opened after four years of construction. This had included the condemnation of houses, churches, businesses, and even a hospital.

There were never any plans to put trains or subways on the Verrazzano-Narrows Bridge. Many traffic experts and citizens have criticized this fact and continue to do so. Licenseless Robert Moses was a rabid fan of automobiles even though he never drove or owned a car. He saw the automobile as the future of transportation, so most of his projects relied solely on accommodating cars—even though it was a proven fact, going back to the opening of the Long Island Expressway, that if you build highways, cars will come, and the roadways will be plagued by traffic tie-ups.

New toll booths at the Goethals and Bayonne Bridges were also introduced that day. The Verrazzano-Narrows Bridge would eventually connect by a series of roadways to the three bridges that crossed to New Jersey. These roadways would flow into what was originally called the Clove Lake Expressway. By the time it opened, it was called the Staten Island Expressway.

Chapter 7

CONNECTING HIGHWAYS AND ENVIRONMENTAL CHALLENGES

By 1963, the cost of building the Staten Island Expressway had reached $1.6 million. It was expected that the expressway would be completed a few weeks before the Verrazzano-Narrows Bridge officially opened. According to Borough President Maniscalco, the bridge would "bring an arterial parkway system" that would make it easier to traverse the borough. This system would make the outlying sections of Staten Island more desirable. Maniscalco believed that many island roads were simply cow paths. They would soon be forgotten once these modern parkways were built.

The western route of the bridge would spill into the Staten Island Expressway—after a toll payment, of course. Building the expressway was done in sections, with the contracts awarded to different companies. It took more than three years to complete the job.

Construction of the stretch between Concord and Castleton Corners was awarded to a Pennsylvania company. This one-and-a-half-mile section was to have been completed by March 1964, but the company was unable to finish the job; in fact, it filed for bankruptcy in September 1963.[39] The Slattery Contracting Company finished the partially completed section. Slattery was awarded a contract in January 1963 to build 1.83 miles of the expressway, 4.89 miles of access roads between Willowbrook and Castleton Corners, and the 1.44-mile Willowbrook Expressway spur from the Bayonne Bridge to Willowbrook. Today, the Willowbrook Expressway is called the Martin Luther King Expressway. Another expressway contract went to Slattery to complete the portion between Richmond Road in Concord and the bridge.[40]

The Verrazzano-Narrows Bridge as it leads into the Staten Island Expressway, after 1968. *Photo by Jet Lowe. Library of Congress.*

Slattery, in business for forty years, had a reputation for completing jobs in a timely and thorough manner.

At one point Slattery had five hundred employees on the highway project between Willowbrook and Fort Wadsworth. Several hundred more Slattery employees labored building the bridge during its construction. Slattery made roads and drainage systems and reclaimed land in and around New York City.[41] Most importantly the Staten Island Expressway was completed in time for the opening of the Verrazzano-Narrows Bridge.

Approximately 1¾ miles east of the Goethals Bridge, there is an exit from the expressway that leads to what was originally called the West Shore Expressway or Route 440 southbound. It was renamed and is now the Pearl Harbor Memorial Expressway. After many delays, the West Shore Expressway was finished in 1976. Its installation bisected Rossville, and it especially harmed the community of Sandy Ground, where many homes and business were demolished for the roadway. The former West Shore Expressway connects at its southern end to the Korean Veterans Parkway at Charleston, where motorists can either head east to Richmond Avenue or go west on Route 440 over the Outerbridge Crossing into New Jersey. This is the road formerly known as the Richmond Parkway, and more often than not, it is still referred to by this name. It has been a source of conflict for decades.

The original Moses proposal called for the Richmond Parkway to run for nine miles from the Outerbridge to the former Staten Island Community College (now the Michael J. Petrides School) in Sunnyside. Construction was to be completed within four to five years.[42] By September 1964, two contracts between the City of New York and construction companies had led to the demolition of seventy-eight buildings between the Outerbridge Crossing and Richmond Avenue in Green Ridge. A third contract was to remove twelve buildings in the vicinity of Bloomingdale Road, Huguenot Avenue, Jefferson Boulevard, and Richmond Avenue. The Richmond Parkway from the Outerbridge Crossing to Greenridge was completed. The section from Greenridge to the Staten Island Expressway at Sunnyside was not because the plan ran rampant across what we now call the Greenbelt. Mayor John Lindsay, realizing the environmental importance of this land—thanks to several island environmental activists—halted the construction in 1966.

The Staten Island Greenbelt Natural Areas League, or SIGNAL, was the first environmental group to protest the ecological devastation that would occur if the Richmond Parkway was built through the Greenbelt. The fight was continued by Protectors of Pine Oak Woods, a land conservation organization on Staten Island that still strictly monitors this proposed roadway. While the parkway is still officially mapped, there has not been a serious proposal to begin construction since 1970. Unfortunately, there are borough residents who believe that the Richmond Parkway would solve local traffic woes. Of course, if constructed, the parkway would cleave a path of destruction through the Greenbelt, an area that consists of properties administered by several groups. Some but not all of the land is preserved. These tracts include LaTourette Park, High Rock Park, Chapin Woods, Saint Francis Friary Woodlands, the Richmond County Country Club Golf Course, Willowbrook Park, the William T. Davis Wildlife Refuge, and the Bloodroot Valley. The properties are owned by the City or State of New York. In the future, environmentalists would like to see Manor Park and sections of the Farm Colony and Sea View Home and Rehabilitation Center officially added to the Greenbelt. Camp Kaufman, the Jewish Community Center, and other rural parcels are privately owned, but they enhance the natural aspects of the Greenbelt. More than anything, environmentalists hope that the unbuilt sections of the Richmond Parkway will be de-mapped and that land will be included in the Greenbelt. Protectors of Pine Oak Woods has waited for more than fifty years. The highway interchanges or "ramps to nowhere" that existed for many years at what is now the Michael J. Petrides School in Sunnyside are now gone. They were finally removed in 2012.

The establishment of High Rock Park is part one of the Greenbelt story. It is also part of the domino effect of land being threatened by development after the opening of the bridge. High Rock Park Conservation Center now consists of ninety-four acres of open green space in Egbertville and New Dorp. It was on January 10, 1964, that Greta Moulton, an aide at the New Dorp Girl Scout site known as Camp High Rock, went before the Staten Island Citizens Planning Committee to call for the preservation of this unique property. The camp consisted of approximately fifty acres at the time, and the Girl Scout Council of Greater New York wanted to sell it. Instead, Moulton proposed that it be bought by the City of New York for inclusion in the park system. Later that year, the Girl Scout Council sold the site to a developer for more than $1 million. The developer also purchased property that bordered the camp. This land would be known as the Walker tract. It was rumored that the new owners wanted to build on the property. They did, and foundations for garden apartments were quickly laid. Company officials believed if the cement was poured, it would legally prevent any interference with their construction plans. It was at this time that Albert Maniscalco used the term *Greenbelt* in conjunction with the fight to save High Rock. In order to preserve the land, it needed to become New York City property. Bypassing Parks Commissioner Newbold Morris, the protestors approached Robert Moses, who wanted High Rock Park owing to its proximity to his planned Richmond Parkway. Moses was always keen on having natural settings along his roadways. He saw High Rock as a fine edge addition to his parkway. Moses told Maniscalco that if he could get a quarter of the money to buy the land from the City of New York, Moses would secure the remaining funds from Laurence Rockefeller, who headed the New York State Council of Parks.

The developer sold the High Rock property to the city for $1.26 million. The eleven-acre Walker tract would be purchased for $295,000. On July 7, 1965, High Rock Park was officially dedicated. It was determined that High Rock Park would be a unique educational venture. Instead of swings, seesaws, and ball fields, High Rock would provide opportunities for individuals and groups to study nature. Here was the first seed of the Greenbelt.

With the former Girl Scout Camp saved, two gentlemen, Robert Hagenhofer and Bradford Greene, began questioning the need for the highway that would butt up against this lovely natural space. In addition, they noted that the Richmond Parkway would cut through a beautiful swath of green that existed in central Staten Island. Robert Moses refused to change the parkway's path, but he did respond to the argument proposed by environmentalists in the *New York Times*. According to Moses, the proposal would cost $11 million more. He further declared that it had "hopeless defects" and that it was "impractical." Still, he was not done. Moses proclaimed that a remapping of that section would be "impossible to finance, difficult to construct and maintain and impossible to police."[43] Since it was in the *New York Times*, his reply gave protestors a voice throughout the city. Mayor Robert Wagner had previously let bids for the construction of the Richmond Parkway, but the man who replaced him rescinded Wagner's construction consent. He was John Lindsay. It was 1966.

The original route of the proposed Richmond Parkway was to traverse a ridge. It would be approximately five miles long, and it would cleave the Greenbelt in two. Trees, plants, wildlife habitat, and more would be wiped out in order to please a few individuals who mistakenly believed the parkway would alleviate traffic problems.

On January 29, 1966, Parks Commissioner Thomas Hoving joined one hundred people for a hike through the Greenbelt. Ankle-deep in snow in a soul-shivering eighteen degrees Fahrenheit, Hoving clearly stated he was there as an "observer." The group met at Staten Island Academy on Todt Hill Road and proceeded to High Rock Park. Protestors against the parkway's route included members of the Staten Island Planning Committee who were determined to fight for the creation of an alternate route to preserve the environment. Congressman John Murphy spoke to the group at the beginning of the hike. He said, "It would be a mistake to change for the sake of change. The present route is the most feasible and has been approved by the State Department of Public Works."[44] By this time, alternate routes for the Richmond Parkway were being developed.

Landscape architect Brad Greene showed Hoving and other participants where trees, a glacial pond, wildlife habitats, and more would be destroyed if the Parkway went through as planned. Greene was a major influence in attempting to make a sound plan for the development of Staten Island. His work toward saving the Greenbelt led to his being deemed Father of the Greenbelt. With a passion for historic and environmental preservation, Greene spent countless hours in leading the march toward saving Sailors' Snug Harbor and the Alice Austen House and in improving Historic Richmond Town. Brad was a founding member of the Staten Island Citizens Planning Committee, a group that recognized that the radical changes occurring on the island needed to be reined in.

One of the Richmond Parkway's "sister-roads" was never built. Originally, the Shore Parkway or the Shore Front Drive was to parallel the waterline from Tottenville to Fort Wadsworth. A later incarnation proposed that it race around the entire island, creating a roadway ring to speed motorists. Robert Moses stated that the Shorefront Boulevard, as he called it, would "preserve forever one of the most dramatic and unspoiled waterfront sections of the city" while also "furnish[ing] protection against beach erosion and tidal floods."[45]

Robert Moses also envisioned an additional length of the Willowbrook Parkway. It would be installed near the entrance of the former College of Staten Island in Sunnyside and run to Hylan Boulevard at Buffalo Street in Great Kills. The roadway would connect with the previously mentioned Shore Front Drive. This section of the Willowbrook Parkway was considered a long-range project with low priority. Neither plan was ever realized. Even so, this Willowbrook Parkway extension still appears on maps as of 2025. The land that was designated exists in a natural state today.

In order to handle increased traffic from the opening of the bridge, improvements for existing roads were suggested. All the proposed and finished highways were desired by developers. They were aware that land and house values had increased along the Staten Island Expressway. This increase in house values along a highway took many by surprise. Historically, the presence of a highway would be a deterrent to buying or selling a house, but commuters preferred easy access to the highway.

Businessmen and politicians saw the bridge as opening Staten Island to the better of two worlds. Firstly, they believed there would be an infusion of industrial activity. Secondly, the island would continue to serve as a bedroom community for Manhattan. It was believed that the decaying piers on the East Shore would be rebuilt and the waterfront would be reinvigorated owing to the bridge. New factories would replace the abandoned industrial sites of Mariner's Harbor and Stapleton. Because of its central location in the five boroughs, the tristate area, and the mid-Atlantic, state politicians and businessmen proclaimed that the island would become a hub for distributing goods to places far away.

Borough President Maniscalco predicted that the suburban flavor of the borough would be preserved despite the development and redevelopment of retail areas, industry, and port facilities. Maniscalco strongly believed that within five years, the bridge would allow the borough to become the "heart of the city's port activities." In 1964, Maniscalco did acknowledge that headaches and hardships would be part of Staten Island's future. School shortages, overdevelopment, zoning problems, lack of transportation, and a lack of sanitary and storm sewers would soon be plaguing Staten Island and its residents. But these specific problems were not mentioned by Maniscalco at this time.

Relative to the construction of the bridge and the resulting increase in traffic, the Port Authority had set aside $13 million for "expansion and modernization" projects at the Bayonne Bridge, Goethals Bridge, and Outerbridge Crossing. The building of arterial routes to accommodate this increase on the New Jersey side was well underway in 1963, but it was far behind schedule.

When the bridge opened, connecting work at the Goethals Bridge and the Outerbridge Crossing was not yet complete. Major traffic jams spilled into the streets of Elizabeth, Perth Amboy, and even Bayonne. The three bridges were greatly affected by the upsurge in traffic. Usage of the Goethals would double seven months after the opening of the bridge, while volume on the Outerbridge Crossing rose by 17.3 percent.[46]

Port Authority officials reported in the spring of 1964 that they were prepared for the onslaught of additional traffic. They claimed a modern six-lane toll plaza would be finished at the Bayonne Bridge by the end of the year. If necessary, the toll booths could eventually be increased to a dozen.

At the Goethals, a modern toll plaza that held administrative offices was planned for the Staten Island side. The offices would span the tops of the toll booths, making for a majestic entrance. The toll booth plaza and bridge roadway would eventually connect to the Staten Island Expressway after its completion. Later, it would join with the future West Shore Expressway.

Viaduct approaches at the Outerbridge Crossing were widened for the rise in traffic. This work was expected to be completed in the fall of 1965. An up-to-date toll plaza on the Staten Island side would also be constructed to coincide with the installation of the Richmond Parkway and the West Shore Expressway.

By 2009, the bridge was generating an astounding $1 million every twenty-four hours for the Triborough Bridge and Tunnel Authority as two hundred thousand cars crossed the span daily.[47]

In 2014, Chairman Thomas Prendergast of the Metropolitan Transit Authority (MTA), the governing body of the Verrazzano-Narrows Bridge by this time, emphatically stated, "The MTA is keeping its promise to ensure fare and toll increases are as low as possible and these options are designed to minimize their impact on our customers."[48] Island residents resent toll increases for several reasons but most importantly because the Verrazzano-

The Outerbridge Crossing between Staten Island and Perth Amboy, New Jersey, circa 1930–1950s. *New York Public Library.*

Narrows Bridge is the only access road for a person driving a vehicle directly to the rest of New York City and New York State, if they stay within the borders of the city or state. Public hearings are held for residents to voice concern about proposed toll hikes. Oddly enough, the public rarely attends. A toll hike hearing was held at the centrally located College of Staten Island on December 10, 2014.[49] It did snow that evening, which contributed to the sparse attendance. Councilman Vincent Ignizio announced that the bridge was a "cash cow" for the MTA. He complained that the MTA's financial stockpile was never spent for improvements relative to Staten Island. As of January 18, 2025, passenger vehicles were charged $11.19 in one direction. The charge for a one-way trip with E-ZPass was $6.94.[50]

Toll hikes always strengthen the argument for improved mass transit on the island. Many residents complain that buses are unreliable and too slow for their needs, so they continue to rely on their automobiles. Still there is no subway on Staten Island, and there probably never will be. Of course, the Staten Island Rapid Transit is available, but it is limited. Requests for a North Shore rail line to be reinstated go unanswered.

Owing to the ever-increasing population of the borough, the opening of the bridge, the upsurge in house building and accompanying roads, and a modern era in which most residents wanted and did purchase one if not two automobiles, both traffic and parking became difficult for residents and visitors. Both are still difficult today. It must be kept in mind that some island roads were originally Native American paths. These same trails could barely be modernized as they would encroach on private and commercial properties. Thus, the expansion of many sections of Amboy Road, Arthur Kill Road, Richmond Avenue, and Richmond Road was prevented. Even so, they are considered main thoroughfares.

Because Richmond Avenue in Graniteville, Bulls Head, and New Springville was still surrounded by farmland into the 1960s, it was possible to convert sections of this one-lane road in each direction into four-, six-, and even eight-lane roadways at various locations. Can anyone imagine that the corner of Merrill and Richmond Avenues had a stop sign in the 1960s? Traffic did arrive, and it arrived quickly after 1964. Hundreds of signs and traffic lights were soon needed in this strange new world.

As officials began to notice traffic flow problems on the "major" roads of the borough, some experimental changes were made in the timing of traffic signals. One trial run of automating traffic lights on Hylan Boulevard occurred in 1965; the change soon became permanent.

Additional parking lots were also needed and installed at larger retail areas of the island. By late May 1963, two lots—one on Park Avenue, between Castleton and Anderson Avenues—were servicing Port Richmond visitors and shoppers. Another lot was to be installed on Taylor Street south of Castleton Avenue in West New Brighton. The Site Selection Board also scheduled a hearing at city hall during the fall of 1964 to discuss a proposed municipal lot at New Dorp Lane and Fourth Street. The site would accommodate forty-two cars. It was approved. The Castleton Corners Board of Trade met to ask the city for a municipal parking lot in anticipation of the contemporary post office scheduled to open on Manor Road in Sunnyside. In mid-November 1964, the group described the current parking issues as "acute." The request was revisited the following summer by Borough Hall.

The opening of the Verrazzano-Narrows Bridge aided in changing the public's shopping habits. While "Main Street" shopping was being replaced across America, the bridge assisted in ushering in the popularity of plazas and malls that catered to car owners. It was not long before small businesses on Main Streets in Stapleton, Port Richmond, West New Brighton, and New Brighton began feeling the effects of the large shopping centers that were opening across the borough. One Stapleton hardware store owner succinctly stated in 1965, "We need to attract our people back."[51] He noted the allure of being able to buy multiple types of products in one store visit and observed that shoppers could purchase food, clothing, and household items at the same time by visiting modern retail centers like Korvettes. A fish market owner stated that the small merchants of Stapleton were doing well until the large stores opened. The manager of Woolworth's on Water Street suggested making part of Tappen Park into a parking lot. He thought a catchy name for the new accommodation would be Tappen Parking Lot.

Stapleton store owners also noted that recently relocated residents to the island were not moving to older neighborhoods; they were relocating to places like New Springville, Willowbrook, Bulls Head, and Eltingville. Staten Islanders, both new and native, were focusing on central Staten Island and parts of the South Shore. Modern homes, businesses and shopping centers were constructed in these areas. The popularity of older North and East Shore neighborhoods was dimming.

This left older established communities nearly deserted. It was a crisis situation for many small businesses—most of which were family owned. A new parking area in Stapleton did not help merchants. By 1970, the stores were dying, and they eventually closed. Referring to the situation as a "Stapleton Shopping Crisis," thirty members of the Stapleton Board of Trade met with representatives of the City Planning Commission, the New York City Traffic Department, and the Office of Staten Island Development in order to come up with ideas to save the merchants. The meeting focused on how to bring shoppers back to their stores. Their sales were being pummeled by the success of modern stores and shopping areas with super-sized parking lots. How would they deal with island residents who were fast becoming more accustomed to using their cars instead of their feet?

Chapter 8

"JERRY-BUILT, RUBBER-STAMP ROW HOUSES SPRING UP IN THE SYLVAN WILDERNESS"

While Main Street shopping was finding it difficult to survive, the business of building houses on Staten Island was thriving in the early 1960s and onward.[52] The *New York Times* made two succinct statements in its January 22, 1966 editorial "Staten Island Down the Drain." The first assertion read, "The sack of Staten Island continues as if it were preordained. Nobody seems to be able to turn back the invaders." The second declaration: "The bulldozers are, every day, destroying wooded hills as matchstick houses march ruthlessly across fields and farms." Someone was noticing the ruin of Staten Island. This newspaper continually focused on the destruction of the borough.

As construction of the Verrazzano-Narrows Bridge advanced, so too did the hoarding of real estate in the "Forgotten Borough." In 1963, eight of every nine residents were located in a one- or two-family house. The borough consisted of something akin to small villages in its West Shore, Mid-Island, and South Shore areas. There were only 7,000 apartment units in existence. Truck farms still occupied 1,100 acres. By 1963, real estate investments had multiplied by as much as 15 percent for land speculators. In 1953, 1,079 building permits were issued. In 1962, 1,642 permits were distributed.[53] Statistics from the New York State Labor Department show that in 1964, approximately 469 island contractors had a total payroll of $19,661,420.[54] Even so, Staten Island of the 1960s was a rural location. The public still attended professional rodeos. On October 17, 1965, the Sheriff's Mounted Posse of Richmond County sponsored a rodeo featuring competitions in bronco riding, steer wrestling, calf roping, and barrel racing at the beloved Weissglass Stadium. Cowboy clowns were also on hand to provide farcical entertainment. Professional cowboys vied for cash prizes while attendees were invited to physically participate. The Sheriff's Mounted Posse, the Richmond County Posse-ettes, and several bands kicked the festivities off with a parade in Port Richmond. When the rodeo was not operating, Weissglass Stadium hosted popular stock car races. While the cars raced and the Posse-ettes paraded through the pastoral neighborhoods of the island, much of it was being devoured.

Former farmland now featured housing developments. Cheerfully colored plastic pennants were strung along the recently laid concrete lanes of Staten Island's contemporary "neighborhoods." Fascinating names were given to these communities. Greentree Estates at Oakwood featured models launched with such exquisite names as the Fleetwood, the De Ville, and the Bonneville. Shore Haven Terrace Homes went up in Great Kills. Mayberry Heights in Bay Terrace was proud to showcase its model known as the Regency. University Heights at Slosson Avenue in Castleton Corners was home to a model home dubbed Athena. (There was no university nearby.) Brentwood was a contemporary development at Signs Road in New Springville. One could visit a model designated the Skylark. Electra II was erected at Mountain View Avenue in Meiers Corners. It was described as "Staten Island's most beautiful Total Electric Model Home." (It has since been converted to include gas.)

Those interested in buying a borough home were invited to meet Barbara Ann Gloede, a beauty queen competitor from Great Kills who was also Miss New York State. Miss Gloede would be competing in the 1963 Miss America Contest. In January 1964, she greeted potential buyers at a two-family model located at 237 Giffords Lane in Great Kills.

For years after the opening of the bridge, a parade of articles and advertisements appeared regularly in the *Advance* announcing new building projects on the island. Housing developments, individual houses, and businesses were all featured. Other articles noted which individuals or companies had submitted plans to the Buildings Department for new construction. For example: March 7, 1964, featured the announcement that "11 1-family houses slated for 2 areas" were to be built in West New Brighton and Port Richmond. A similar article the previous month headlined the construction of "22 homes to be built in Graniteville." Another twenty-two one-family houses were to be built at locations around New Dorp. The same article documented other one- and two-family houses at Great Kills, Graniteville, Pleasant Plains, Annadale, Eltingville, Meiers Corners, Tottenville, Richmond (Town), Midland Beach, Dongan Hills, and elsewhere. Central and southern Staten Island were under assault. It would not stop.

All the plans submitted to the Building Department meant the arrival of money through fees for the City of New York. In March 1965, the entity received more than $22,000 from applications for new construction.[55] Adjusting for inflation, that amount would equal over $196,000 today.[56]

There was certainly much to attract newcomers in the 1960s, especially those who lived in congested, overbuilt Brooklyn. They could enjoy detached or semi-attached affordable housing as opposed to multifamily houses or apartment buildings. Staten Island houses cost less, and taxes were lower. There was ample parkland. Five thousand Staten Island acres were designated as such, and $3 million had been earmarked to improve parks, playgrounds, and beaches. There were still nineteen thousand unbuilt acres across the island.[57]

Indeed, in the spring of 1964, house building continued to be the largest industry on Staten Island. When the House Builders Association publicly noted that "everyone living in the borough, all business, commerce and financial institutions are directly affected by the building industry," it was speaking the truth. The association vowed, "It is an industry that is going to get much bigger and even more vital to our community's life—and soon."[58] The House Builders Association also contemplated: Where would people live when the population reached five hundred thousand? Where would children go to school? How would people get around the island's fifty-seven square miles? Where would they shop and play? No answers were provided, other than the statement that the building industry must be involved in solving future community problems.[59]

According to the association, it was in "constant touch with city officials. When unreasonable and unnecessary regulations hamper and delay growth and progress on Staten Island this relationship is essential to help remove frustrating bottle-necks."[60] It was asserted that the building industry was intensively concerned with mapping and sewers, but the building went on whether maps were complete and whether a house was immediately connected to a sewer or not.[61] Buyers were advised to consider their needs for both the present and the future. The association recommended its members as they were responsible and reputable. It stressed that a code of ethics had been adopted by the group. Its "first pledge [was] 'to promote the welfare of our community, our customers, and our country.'"[62] Members also guaranteed the "operation of plumbing, heating, ventilating, and against foundation water seepage or faulty roofing, flashing, or other openings in the building" for one year.[63]

During March 1965, builders accused the New York City Highway, Public Works, Health, and Building Departments of "abuses of authority and complete lack of cooperation."[64] They claimed these abuses "cost the building industry of Staten Island undue delay and great expense and threaten[ed] future developments."[65] In response, the deputy mayor requested that the borough president obtain evidence of the alleged abuses.

Banks were making money, too. One bank president announced in 1964 that there "was 'plenty of mortgage money' available and that there were few, if any, foreclosures." He maintained that "55% of the mortgages now granted are being obtained by off Staten Island families."[66] The increase in population and housing led the commissioner of the Department of Buildings to state that the opening of the bridge enabled Staten Island to become the fastest-growing borough in the city. His department documented that between 1961 and 1964, the estimated cost of a recently built house rose to $31,176. In 1961, that cost was approximately $15,757.[67] The cost of a house had doubled in three years.

By 1964, builders that once put up twelve houses a year were building fifty to one hundred. Asking prices were always accepted owing to the demand. As soon as another house was built, it was bought. Natives quickly perceived that the bridge was ending a way of life, one that kept them isolated from urban New York City. The other four boroughs were no longer at a safe distance. The natives projected the arrival of overcrowded schools, inadequate roads, and intrusive highways, and the loss of scenery, open space, and above all, privacy. Their perceptions were correct. The public was queried by the *Advance* on their views regarding the preplanning that should have occurred before the arrival of so many newcomers relative to the bridge. Anonymous responses published on October 5, 1965, railed against the lack of planning and blamed the city for attempting to plan in hindsight. The lack of streets, sewage, water mains, and schools were cited by one individual; another complained that the recently built houses were too close together and they all looked alike. The general feeling was:

> *The hodge-podge zoning ordinances now in effect...are the greatest single detriment to the orderly and proper development of Staten Island. They should be examined and changed by the best minds of Staten Island's own professional and business experts, not by outside, disinterested engineers.*[68]

One surprising lecture on planning and building came from Robert Moses, the catalyst behind the entire recasting of Staten Island. On April 13, 1957, he frankly stated that the bridge would bring families seeking "modern low-and-middle-income housing." According to Moses,

> *The big thing is to avoid the mistakes made in other suburban housing, to curb the selfishness of developers, the lowering of zoning standards, poor construction on postage stamp lots and the disappearance of the countryside. Your uneven topography helps you* [Staten Islanders]. *The developers love flat land and hate hills and valleys not susceptible to simple cut and fill operations.*[69]

All this was true, and all this did happen, even though borough residents were warned by the man who was bringing the changes. (Neither hills nor valleys stopped developers until citizens began fighting for hillside preservation in the 1980s.) But did Robert Moses really believe that the average citizen could stop the landslide of development that was about to arrive? Most houses were constructed in developments. Natives complained that there was no individuality, that the developments resembled Levittown, New York. Inhabitants screamed that Staten Island was turning into another Queens.

One realtor blamed the public for the poorly built homes. He said builders could only put up what could be sold—in other words, what people could afford. Professional planners responded that builders knew nothing about site planning; they "just meet the minimal standards and make as much as they can" off what they've built.[70] The Staten Island Real Estate Board called for stricter zoning in order for the island to retain its suburban character.

Members of the Home Builders Association, responding to criticism that they were wantonly destroying the borough, claimed that their aim was to provide "decent housing for as many people as possible."[71] Developers claimed to be supportive of all planning efforts, but they were quick to state that house building was the borough's foremost industry. Builders asserted that they "would not want to see their multi-million dollar business grind to a halt" as a result of planning measures that would block economic growth for the borough.[72] According to the Home Builders Association, "The welfare of the people of Staten Island should not be sacrificed on the political altar."[73] Builders complained that they were not invited to participate in planning sessions. They accused planners of believing that they were not qualified to discuss planning issues. Builders took offense at terms such as "ticky-tack" houses and were confident that they did an "excellent job" in providing a "well-built home with maximum value consonant with the price being paid."[74] According to the association, zoning regulations were outmoded and forced builders to create a "grid envelope" design. Builders were also dissatisfied with the city's recent policy of withholding land from public auction, a tactic that officials hoped would slow the building boom. They further hoped to set aside land for better planning in the future. Withholding these plots would also allow for future schools, parks, utilities, and much-needed sewers. Builders suggested this only caused land prices to rise and that the city was playing into the desires of speculators. Builders were accused of disregarding architecture and creating rows of look-alike houses in order to sell the least expensive product possible. But builders claimed that by duplicating and selling a mass-produced product, they were providing what the public wanted. Builders argued that the grid system of streets corralled them into replicating houses. They blamed centralized powers in Manhattan and complained that the borough president was weakened in controlling matters of building and planning on the island. They pointed to neighborhoods like Westerleigh and New Dorp, where laid-out grid styles were still pleasing to the eye.[75]

A 1966 editorial in the *New York Times* entitled "Farewell to SI?" squarely blamed the mayor's office, the Staten Island borough president, the City Planning Commission, and the City Real Estate Department for the borough's lack of planning in a number of key areas. These included pre-mapping, the route of the Richmond Parkway, and the "shameful sell-off of city-owned land for unplanned development"—a.k.a. in rem sales.[76] Politicians admitted that they would not take a stand on legislation that called for pre-mapping before construction. Until 1963, a house could not be built if a street was not mapped, but then the ruling was changed.[77]

Architectural expert Ada Louise Huxtable wrote in the *New York Times* on August 9, 1965, that many spot developments occurred on the island as a "result of the City's two year grace period in implementing the new zoning code of 1961." This "allow[ed] speculative investors to file enough old zoning law blockbusters to glut Manhattan for another generation....It also permitted Staten Island developers to file and build minimal housing on minimal lots outlawed by the new legislation."[78] She characterized the building boom as "back door planning" with "jerry built, rubber stamp row houses that are springing up in the sylvan wilderness like August weeds."[79] Huxtable blamed the limitations of zoning for much of what was occurring and the lack of a master plan. She acknowledged that nothing was being done to "promote the creative design of planned new communities." Huxtable labeled any effort made toward sensible design as "comparable to fighting a forest fire with an eye-dropper."[80]

The following day, the headline of a *New York Times* editorial read "Ruining Staten Island." The editors wrote that the "most rural borough is being mashed by the bulldozer and the speculative builder into an unlovely setting for future slums." The newspaper decried the lack of government involvement in the "despoliation," and they noted the preponderance of poor taste in new construction.[81] The well-heeled of New York City also responded. Elinor Guggenheimer, a wealthy member of the New York City Planning Commission, expressed her support for making Reed's Valley into a park while also improving development of the entire area. Guggenheimer was brought on a tour of this location on October 3, 1966. She was thrilled by the woodland but appalled by the loss of the hillside as well as other changes to the topography that had been made by developers.[82] Various other problems were brought to Guggenheimer's attention, including houses that were built in cut embankments, which caused storm water, mud, rocks, and other debris to cascade onto the property of Emerson Court residents below. This further resulted in undermining yards, shrubbery, and even house foundations. Basement floods were now a common occurrence in the area. Residents even introduced her to a piece of property that had been marred by developers for the sole purposes of digging out fill for their own projects.[83]

There were still people who called for sensible borough planning. In 1963, the Master Plan Group of the Staten Island Citizen's Planning Committee had completed a yearlong study that envisioned a series of planning problems if the population continued to increase as expected. The group was headed up by Terence Benbow, Bradford Greene, and others. The committee called for a strategy that would give the borough "an identity of its own." They called for the "establishment of a general growth pattern via a master plan which would focus attention on such local needs as transportation and parking, street mapping and community interrelationships."[84] The group noted that pedestrians were rarely considered in any projects and this needed to rectified for safety purposes. They called for the preservation of local historic landmarks and for rehabilitating historic homes in declining neighborhoods.

The Citizen's Planning Committee strove to beautify the borough. It presented a program to the Livingston Community Association titled "Eyesores—Evermore or Nevermore" on April 5, 1965. Brad Greene and Robert Hagenhofer stressed that trash across the North Shore needed to be removed. They suggested that property owners clean up and stop carelessly discarding rubbish in convenient but illegal places. A slideshow featuring garbage, abandoned cars, and old refrigerators was viewed. It was noted that illegal dumping was not a modern problem, but there was a sharp upswing in the number of incidents as the drive to build up the borough marched along. Dumping was easy—sanitation personnel worked days, so dumping occurred at night. It also appeared that citizens dumped illegally because they were not aware that the public had free access to drop trash or bulk items at the Fresh Kills Landfill. Regarding contractors and builders, the process for getting the necessary permit for legal dumping was complicated. It involved two applications, three trips into Manhattan, a trip to Brooklyn, and a fee of thirty dollars. It was further noted that the payment was not prorated if the application was not submitted on the first day of the city's fiscal year. With approval, truck owners were allowed twenty-five dump visits per year. The cost for this privilege was fifty dollars for a truck that held five yards of refuse. Many thought that paying the ten-dollar fine if caught dumping illegally was easier than all the runaround and expense.[85]

Brad Greene was a strong proponent of improving the island. He strove to better the environmental conditions of the borough while fighting to save open space for natural purposes. Mary Lou Greene, Brad's spouse, was also a powerhouse of activity who believed in enriching her home borough. Mary Lou was a home economics teacher at Prall Junior High School in West New Brighton. Other nonprofits actively supported by Mary Lou Greene were the Staten Island Museum, the Society for Seamen's Children, the Snug Harbor Cultural Center, and the League of Women Voters, among others.

In 1991, Brad Greene was honored at the seventh annual Greenbelt Gala for his dedication to preserving this and other green spaces. Greene is well remembered for working to preserve and improve High Rock Park, Willowbrook Park, Westerleigh Park, and Clove Lake Park. Brad Greene served on the board of trustees of the Metropolitan Museum of Art and the Staten Island Museum. It was he who designed the decorative gate that now stands at the entrance of High Rock Park.

The spirit of the 1960s called for change. This spirit enforced that the environment and historic structures needed to be safeguarded and preserved. These changes were also spurred on Staten Island by the opening of the Verrazzano-Narrows Bridge and the resulting upheaval it brought.

The Staten Island Museum (as seen in this circa 1960s postcard) was one of many nonprofits supported by Brad and Mary Lou Greene. *New York Public Library.*

In addition, transformation in land use, transportation, housing, education, population, and more was realized. Much of this change was chaotic. Even so, there were some who pressed for an organized strategy.

On January 30, 1965, Wagner College hosted the symposium "Island in Search of a Plan." The meeting brought together numerous individuals and groups interested in a borough-wide master plan. The seminar was organized by the Staten Island Federation of Planning Agencies, the Staten Island Chamber of Commerce, the Staten Island Citizens Planning Committee, New York Telephone, Richmondtown Restoration, the New York City Board of Education, and other entities interested in health, culture, social welfare, and human relations.[86] Many of the participants were passionate about establishing a plan for the betterment of the island, but they were unsure how to proceed. Several were interested in their own plans, which were geared toward their interests or the interest of specific groups that they were affiliated with or represented. These plans and interests, more often than not, coincided with financial gains. They claimed that creating a master plan was dangerous and that privately owned properties would be condemned if a plan was created.[87] Without a master plan, they saw immediate positives that would enhance their profits. These included increased tax revenue, a booming local economy owing to a $30-million-a-year construction industry, and apartment dwellers becoming homeowners thus increasing the number of houses built and sold.[88]

During June 1966, the New York City Planning Commission finally released a plan for Staten Island's future growth and development. Two features were immediately noticeable. Firstly, the blueprint called for a Staten Island development coordinator. Attorney Holt Meyer would be appointed. Secondly, the plan called for the "formation of a quasi-public agency to oversee development of the South Shore."[89] Simple yet important recommendations also included establishment of a timetable for sewers, schools, streets, and paving; stricter controls of city-owned land; zoning changes to permit cluster developments; improving Hylan Boulevard; modernizing the Staten Island Rapid Transit system with the idea of eventually linking it to Manhattan via the Bayonne Bridge; and preserving and expanding the Greenbelt concept.[90]

The report was officially titled *Staten Island Development: Policies, Programs and Priorities*. Holt Meyer was an ally and appointee of Mayor John Lindsay. He had unofficially advised Lindsay about island issues since the commencement of Lindsay's mayoral campaign. Meyer said he was "taking office under an executive directive [that] gave him wide powers in areas affecting development." He was to work in the realms of planning, mapping, zoning, and coordination within the borough. Holt Meyer was to report directly to the mayor. The thirty-five-year-old West New Brighton resident saw his job as "ending development that is haphazard, tasteless, and monotonous and improving its character and standards."[91] Meyer noted that creation of the position was long overdue. He and the City Planning Commission called for the prevention of "the same urban sprawl, urban neglect, and despoliation of natural features" for the city's "Last Frontier."[92] This label was affixed to Staten Island as it offered the final opportunity to preserve and utilize the remaining natural characteristics of New York City. Meyer was also responsible for coordinating city departments, negotiating with builders, suggesting capital projects, and developing procedures for controlling borough expansion. He was to be sworn in on August 4, 1966. Meyer believed he would be settling into his Staten Island office a week or two thereafter. But there was a problem. Borough President Robert Connor was not pleased that the mayor had appointed a development director for the island. Connor made it clear to Lindsay that he would have a "'king-sized fight on his hands' if Meyer's role is aimed at setting up a political organization for the mayor on Staten Island."[93] Connor was further concerned that Meyer's appointment was an "attempt by the mayor to bypass the borough's elected officials."[94] Connor did not attend Holt Meyer's swearing-in.

The standoff between Lindsay and Connor continued. Connor requested a list of Meyer's duties, which he hoped did not usurp his authority. If he received the list and it met his conditions, the borough president said, he would be happy to provide space in Borough Hall for Meyer. The list was not provided, so Meyer moved into 350 Saint Mark's Place in Saint George. He did not get a Borough Hall office. For the rent on Meyer's Saint Mark's Place office to be paid, the lease had to be approved by the Board of Estimate. The five borough presidents sat on the Board of Estimate, and all five were displeased with Meyer's appointment. They believed it was Lindsay's plan to place "little city halls" in each of the boroughs. The city controller and the city council president supported the views of these borough presidents. Connor declared he would not have opposed the development director position if "it had been created within the framework of the commission, not as a branch of the mayor's office."[95]

Holt Meyer's initial plan for Staten Island included a community development study of the Annadale-Huguenot urban renewal area—already a sore point with residents. Meyer called for an environmental survey that would evaluate the natural features of the island in order to preserve open space. Meyer further believed that a geologic study was needed to examine island locations that were suitable for septic tanks. He also sought a development plan for all of South Richmond; a strip park system that would follow Lemon Creek, Richmond Brook, and Sweet Brook; a study of the "Heartlands Commercial-Civic-Industrial Complex"; and a development plan for the shore areas between South Beach and Great Kills Park.[96]

In January 1967, Meyer created an advisory unit that consisted of fourteen individuals. Landscape architect Bradford Greene, a member of the Staten Island Home Builders Association, and Lyn Carver, a past president of the Staten Island branch of the League of Women Voters, were among those included.[97]

As in previous years, the only planning tool available was zoning. Per a 1916 zoning plan for the borough, which was not based on any master plan, most of the island was given an "unrestricted" designation that provided no rules for development.[98] A 1961 rezoning occurred, but it was not related to a plan for the island. Furthermore, property owners needed only to petition the Board of Standards and Appeals to get "changes in zoning or suspend specific requirements." Thus the board had the power to "permit construction of smelting plants in Great Kills if it thinks it proper."[99] For the most part, it was up to the citizenry to fight zoning petitions. Unfortunately, residents did not have the knowledge or skills to combat groups or individuals who thought in terms of dollars and not quality of life. It was forgotten in 1960s Staten Island that zoning regulations were originally created to better communities.

Chapter 9
UNPLANNED CHAOS AND DISORDER CONTINUES

By the mid- to late 1960s, the word *developer* had become a negative one on the island. The Home Builders Association was founded in 1953. Its president stated that the organization was formed to "uplift the industry." By 1965, membership had grown to two hundred. Architects, attorneys, realtors, insurance brokers, and others were also invited to join. Their unofficial motto was "The best possible housing for the greatest number of people, at the lowest possible cost." The group further stated that it was constantly keeping a "sharp eye out for any practices on the part of the various governmental authorities that might result in a higher cost to the home buyer." They believed that private industry knew what was best for the buying public. The organization stated, "Many of the onerous laws and regulations that we have succeeded in defeating through the years have resulted in savings to the home buyer of thousands of dollars. This Association is constantly fighting to allow greater land utilization." The association was quick to note that it sponsored signs at Little League parks.[100]

In 1963, the New York City Buildings Department had received a record-breaking 2,477 permit requests for new building construction on Staten Island. The majority of these structures were one-family houses. The department also had a corresponding application increase for alteration installations, complaints received, and complaints processed.[101]

Staten Island natives were appalled. In the two weeks prior to October 23, 1965, the Buildings Department office at Borough Hall had received 237 applications for additional buildings. This increased the number

Staten Island natives were appalled when brooks, meadows, creeks, and forests disappeared as developments sprang up across the borough. *New York Public Library.*

of applications for the year to 2,917. When the year concluded, 3,586 applications for new construction had been received. It was a record.[102]

In 1966, the construction industry was seeking sympathy. There was a 10 percent slump in the number of building permits filed between January and July that year, compared to the same period the previous year. One builder insisted that "to estimate the losses in dollars and cents is impossible," while another complained, "It's far worse [than the previous year]. We're suffering hard."[103] Developers blamed the lack of money for loans and mortgages. In response, some construction workers were idle while others took pay cuts.

In 1967, members of the New York Builder's Association complained that choice building sites on Staten Island were disappearing. According to the *Advance*, the spokesman for the association was promoting a "'buy now campaign' aimed at portraying the current financial climate for home buying in the New York metropolitan area as 'never better' than this spring."[104] The spokesman stressed that interest rates, taxes, and carrying charges would only go up, so it was important for the consumer to act soon.

The following year, builders still referred to 1967 as a poor, even a disastrous year. The Home Builders Association stated, "The boom we had anticipated from the opening of the Verrazzano-Narrows Bridge never materialized."[105] Builders claimed that construction slowed by 50 percent in 1967 compared to 1966. They blamed the poor image of the building industry, severe weather, and city government. Vandalism at construction sites was a problem, as was New York City Health Department interference with installing septic tanks. In fact, builders believed that changes in septic tank regulations were going to cause bankruptcy and land loss for developers.[106] They blamed the sameness of development housing on economics, claiming it cost less to build houses in the same or similar styles. They argued that homes needed to be affordable for Brooklyn residents, who were mainly apartment dwellers without a home to sell to meet the need for a house down payment.

Builders were also aware that they could increase their profit while meeting zoning requirements by building duplexes on the standard lot size of fifty by one hundred feet. These duplexes provided two dwellings separated by a fire wall. Staten Island natives might not buy a duplex, but Brooklynites escaping to the island devoured them. (Some people later developed an aversion to them as owners painted either side of the exterior in different colors or used mismatched materials on the façade.) Most customers living on or moving to Staten Island in the 1960s wanted a high ranch. These residences had a basement on a cement slab with a three-room apartment and a garage. A six-room living space was overhead. Given the ever-present threat of floods, it was wise that these homes lacked basements. This was also a major savings for the builder, although it caused the apartments to be ice-cold in the winter. The rental unit was an important selling point since it could cover mortgage payments. The homeowner could also put their mother or mother-in-law in the apartment, hence the term *mother/daughter house*.

By 1967, Mayor Lindsay was disgusted with overdevelopment on Staten Island. He was quoted as claiming that "clogged, formless communities without beauty and order and without regard for people or land" were being put up on the island.[107] As such, it was requested that builders include curvilinear streets and parks in their developments. They were asked not to build over scenic streams. It was further observed that it was dangerous to build over a waterway. In one instance, a builder was asked to donate property for a school in exchange for putting in his own sewage treatment plant within the development. He agreed, but when asked to create a more attractive street pattern, he refused.

According to the *New York Times*, between 1960 and 1967, more than twelve thousand homes were built on Staten Island—with minimal city planning and maximum exploitation.[108] Bulldozers obliterated hills, trees, streams, and any other natural features that were in the way. "Builders covered the denuded land with identical little boxes crammed onto lots."[109] Even so, Brooklyn residents flocked to the borough and loved it. They found there a country atmosphere, though they did have complaints about crowded schools, erratic bus schedules, distant shopping, lack of parks, malfunctioning septic systems, and more—much more.

By 1967, builders were attempting to improve their image. In one article, a builder stated that many building contracts were "written to include a buyer's request that trees be taken down." He continued, "The buyer usually is the owner of a second home who is afraid lot grading will either choke a trunk with extra earth or take soil from the roots, killing the tree." The builder believed that the cost of removing a dead tree from the homeowner's yard was between $100 and $150. "Everybody wants trees, he said, "except the people who are buying a house."[110]

John Lindsay was not a popular mayor. Even so, he was opposed to overdevelopment and the rambling ruin of the island. He did not want the borough to be another bedroom community for the rest of the city. Lindsay strove for "economic diversity—and self-sufficiency" for Richmond County.[111] It was Lindsay's hope and the hope of his urban planners that Staten Island residents could live, work, shop, and enjoy recreation in the same borough. He and Holt Meyer wanted an "ideal suburb" that included light industry, research labs, and small manufacturers to replace the construction business as the most important base of the island's economy. Waterfront development with a marine terminal for containerized cargo was in the works. It was still hoped that the East Shore piers could be revitalized. The establishment of industrial parks was also being investigated.

Even so, the march of development continued. On July 7, 1968, the *Advance* reported that 19,965 dwelling units had been built on the island in the ten years prior to June 1968. As of January 1, 1968, the borough had 82,779 housing units. Of this number, 55 percent were one-family homes, while 29 percent were two-family homes. The remainder of the population lived in apartment buildings or multifamily houses.

Chapter 10

NO MAPS, NO ROADS, NO SEWERS, NO SCHOOLS—NO PROBLEM

The arrival of the "South Shore Mushroom" was proclaimed by the *Advance* on May 4, 1959. Cape Cods and ranch homes from Grasmere to Tottenville to Graniteville were selling quickly. This was occurring throughout the island. Sewers, schools (it was widely believed that a system of junior high schools would prevent school overpopulation), roadways, and more were slowly, very slowly, being prepared for the onslaught of additional residents in anticipation of the bridge.

The Home Builders Association "pledged to the concept that 'our paramount responsibility is to customers, our community and our country,' [and it was] in the forefront of the community efforts to preserve many of the historic features of Staten Island which are distinctive and which all of us treasure." This was in 1964. The association believed that Staten Island had been treated as a "stepchild." "Better government and better living facilities" were said to be among its concerns. Members swore they were part of a special program that vowed a "pledge of integrity" to new homebuyers. The guarantee also promised homebuyers that they would "receive exactly what they pay for."[112]

Borough President Albert Maniscalco announced in 1964 that Staten Island was about to experience an "emergence from its former insular and sheltered life into the limelight of full attachment and participation in the affairs of New York City." Maniscalco hoped an electronic research center on the island would be sponsored by NASA. He believed that what is today Historic Richmond Town would progress as a foremost cultural attraction.[113] The effort to expand Richmond Town was originally supported by Robert Moses.

In 1959, only one contemporary elementary school was planned for construction in the borough. It was a replacement for Public School No. 3 in Pleasant Plains. Residents were hopeful that another high school would be built in the vicinity of Annadale and that modernization at New Dorp High School would alleviate overcrowding. By 1963, nine public schools were being built or plans for their construction were moving along. That might sound promising in terms of student overpopulation, but it was less than what was needed. Officials saw the situation remedied by portable classrooms. They were installed at several of the island's forty-three public schools, which held a total of thirty-three thousand students at the time. Owing to the increasing population, many schools were on a double session. Fifty private and parochial schools, which held twenty-thousand students total, alleviated some of the public school overcrowding. This situation was occurring even before the bridge opened in 1964.

The New York City Transit Authority had done little to improve bus service in anticipation of the increasing borough population—even though buses were commonly crammed full during high commuter hours. Of course, there was no subway or light rail planned to cross the bridge or Staten Island other than what was already in place. In fact, Councilman Robert Lindsay described the existing mass transit system on the island in 1965 as "good enough for the horse and buggy days, but far from in line with the anticipated population explosion and even the borough's current needs."[114] As such, Lindsay directed the Planning Commission to conduct a study that would provide recommendations for improving the system as soon as possible. At a time when most municipalities were advancing and expanding mass transit, Staten Island was only concerned with bridges, highways, parkways, and roads. Between 1960 and 1966, the number of motor vehicles on Staten Island jumped by approximately 38 percent. As islanders knew and as arriving Brooklynites soon realized, you needed a car to get around the island. The Planning Commission called for reactivating the North Shore rail line and reorganizing island bus routes to keep up with commuters and residents. The commission further urged that mass transit from Staten Island be directed via the Bayonne Bridge to Manhattan. Neither suggestion would be seriously considered at that time, and both ideas are still bandied about in the twenty-first century.

Staten Island was viewed as an "explosively developing strip" between Boston and Washington, D.C. It was isolated, incompletely mapped, and

The Bayonne Bridge was one of three crossings to New Jersey that experienced an upsurge in traffic due to the opening of the Verrazzano-Narrows Bridge. *Undated photo, New York Public Library.*

still rural—a surprise to many as it was part of New York City. Albert Maniscalco was borough president as the Verrazzano-Narrows Bridge went from drawing board to reality. He served from 1955 until 1965. That last year, he noted that the most important influencing factor for Staten Island's growth was the lack of a master plan. He said, "There is no unified set of rules to control the outcome of the boom."[115] According to the borough president, there was never enough money to develop a master plan since New York City would provide only $3,000 to complete the project. In fact, no plan had been developed for Staten Island even though creating one had been mandated.[116] To be fair, the 1961 revision of the New York City Charter took much power away from the borough presidents when it went into effect. Street construction, street repair and maintenance, and sewer installation had been important functions of Borough Hall. Now they were the responsibility of municipal offices in Manhattan, so borough presidents needed a good relationship with city officials. In 1966, a power shift occurred when Robert Connor became borough president while John Lindsay was mayor. They were both Republicans in a city that screamed allegiance to the Democratic Party. Theirs was a pleasant relationship at the start, but it quickly became disagreeable. This situation was one of several that negatively affected the changes occurring on Staten Island.

Albert Maniscalco was able to prioritize a mapping project in 1960, but it took years to finish. Before its completion, builders started building, and they did not stop. By 1965, much of the island was still not officially mapped. Even so, building permits were distributed. On March 1, 1966, the Staten Island Chamber of Commerce urgently requested mapping streets be made a priority. The chamber stated that half of the borough's streets were not official streets because they had not been mapped as required. "Mapped streets, the Chamber reason[ed], would solve the Building Department's current difficulty of having to withhold certificates of occupancy because streets on which the builder has erected and sold new homes are not technically final."[117] In the fall of 1966, it was reported that more than 1,800 homes were "illegal" because they lacked a certificate of occupancy.[118] The certificate ensures that minimum building standards have been met and describes how the building can be used: two-family house, store, and so on. Not having a "C of O" was problematic for a number of reasons, including that the certificate was needed to resell a building.

Staten Island had 700 miles of mapped streets in 1955, only 462 of which were paved. A miniscule 297 of these streets had sewers.[119] The 238 unpaved miles were referred to as paper streets. These roads were set up in a gridiron style. This pattern was considered old-fashioned and outdated and it did not align with the island's natural topography, hence the call for curvilinear streets. Some of the paper streets that were laid out in the 1920s "phantom street system" allowed for subdivision maps that were never made official.[120]

Suggestions of planning relative to the changes that the bridge would bring did circulate. In the neighborhoods of Annadale and Huguenot, 1,083 acres were to be planned as a coordinated unit. For this reason, city-owned land was officially held back and not sold at auction. But creating a logical plan for the future of Annadale and Huguenot was termed *urban renewal* since this program would be used to finance the plan. Unfortunately, this term frightened most politicians and residents—especially those in Annadale and Huguenot. Local residents imagined a "federally assisted urban renewal bulldozer bearing down on their homes—'Big Brother' at its worst." Urban renewal or not, the bulldozers began tearing the borough apart without the benefit of planning, and neighborhoods were "scarred by pockets of private, speculative mutilation."[121]

Ada Louise Huxtable continued to be appalled as the island was "defaced by builders' dusty, earth-strewn plots" along half built and, even in the midst of New York City, dirt roads. Trees and telephone poles bore the signs of soon-to-be-forgotten development names that offered recently built homes with such descriptions as "distinctively styled split levels and high ranches." Huxtable observed that "single block, dozen house, quick buck projects of the small, short-term operator whose costs and prices are high, whose standards are low and whose product has a scattered, erosive, small shot destructiveness that defaces every stretch of unspoiled land" went up across the island. She proclaimed that hills were "stripped bare and treeless houses march unfeelingly across them in straight bulldozed rows."[122]

The arrival of apartment buildings led many to speculate that urbanization was inevitable. Several, with striking views, were built along Victory Boulevard. Apartments generally rented for fifty dollars per room, but renters were slow in arriving. Brooklyn transplants wanted houses. One apartment building that opened had seventeen stories. Parkview House stood at 700 Victory Boulevard. It held 228 units. Tenants began tenanting in 1963. The tallest residential structure on the island, the building had a pool and sat on a landscaped property that overlooked Silver Lake Park. Each apartment had a terrace. Rents started at $119 per month.[123]

Roads were ripped up. Con Ed was ripping up roads. Brooklyn Union Gas was ripping up roads. The City of New York was ripping up roads. Fuel line companies were ripping up roads. Developers were ripping up roads. And all this ripping was done on a regular basis and usually right after a previous ripping. On tiny Merrill Avenue in Bulls Head, around 1967, a Grand Canyon of sewer pipes was being installed to accommodate new homeowners on Arlene Street, Fieldstone Road, and other neighboring streets. But no hookups were allowed for the embattled septic tank owners of Merrill Avenue. As a matter of fact, the houses that stood there in the 1960s through the 1990s were never connected to city sewers.

Two streets in Willowbrook impaled by utility poles made the local newspaper on September 2, 1966. The poles stood in the middle of Willowbrook Road and Forest Hill Road when builders paved the roads beyond their original width to meet newer standards. In doing so, they simply paved around the intrusive poles and left them in the center of the street. Fire hydrants were also left in the middle of roadways at several locations.

One construction activity that irritated residents was the widening of streets. Sometimes widening occurred after new developments were in place. On August 12, 1965, residents of Mason Avenue in Ocean Breeze were so disgusted that they protested. Their street was to be widened and raised by four feet at their front doors, making it impossible for them to exit their houses. In addition, the New York City Highway Department would be removing the existing sidewalks and front lawns, and homeowners were

Fire hydrant left standing in the street at Lamoka and Fern Avenues, Great Kills, September 9, 1971. *From the* Staten Island Advance.

expected to reinstall both at their own expense. It seems the houses had been built and bought three years prior, but no one had informed recent homeowners that critical changes would be taking place in the future. One resident, Mrs. Leonard Calia, wrote to all her elected officials, including President Lyndon Baines Johnson, who was in fact the only person to reply.[124]

The headline read "Housewives and Babies Protest." Mothers arrived with beach chairs and children to express their outrage over the lack of storm sewers in Dongan Hills. Parking themselves at the corner of Buel and Mason Avenues on August 7, 1967, they vented their anger after spending the previous week removing floodwater from their basements after soaking rains hit the island.[125] Participants also staked out spots within the boundaries of Slater Boulevard and Naughton and Seaver Avenues. Proclaiming that the city had not answered their complaints, they were taking to the streets to alert anyone who would listen. The problem also involved sewage backing up into basements. Protestors held placards that read, "Give us sewers not swamps." Owing to the perilous water conditions that resulted from storms, mothers were concerned that their children would drown as they walked to nearby PS 52. Some of the women blocked intersections, hoping to prevent cement mixers from dropping their loads at construction sites. Others warned prospective buyers to stay away from the defective location. A spokesperson for the group blamed much of the problem on older homes built below the legal street grade and newer homes built at a legal grade that was higher. This caused rainwater to flood streets, yards, and then basements.[126]

Fifteen disgusted mothers and several children were so upset by the flooding that they picketed the Bard Avenue home of Holt Meyer on September 6, 1967. One Naughton Avenue resident told the *Advance* that residents were tired of talk. They wanted action on storm sewers. The resident threatened that the homes of other borough officials would be targeted. While the protest was in full force, Meyer was actually at his office. His wife, Kathy, was in the house. She decided not to bother her husband by telephoning him. Meyer later stated that the demonstration would "have been more meaningful if it had been conducted at my place of business." The residents were desperate for storm sewers as construction and heavy rains made it impossible to avoid flooding. One protest sign read: "Our Children Walk the Plank to School—Help."[127]

The same group took Meyer's advice and rallied outside his Saint George office on September 12, 1967. They were persistent in showing their displeasure with poor drainage. Meyer's staff informed the group that "preparation of the final drainage plans and mapping" was being worked on. Unfortunately, the picketers' request for storm sewers would not be fulfilled anytime too soon as the city needed $25 million for the project. Aside from the money, the area was still being developed, so the storm sewers could not be installed until all construction was complete. In 1967, builders were still constructing houses at the "existing legal grades, which [were] from one to three feet higher in some places than those recommended by [engineers], and as much as 10 feet higher in many places than the existing grades of some streets."[128]

In 1965, builders raised Rochelle Place approximately thirty inches, with the result that homeowners on the west side of this Eltingville street were now facing a thirty-inch wall.[129] The unwanted wall made driveways useless. The builder promised ramps from the street to the driveways, but they were slow in arriving.

Meanwhile, homeowners on Rensselaer Avenue in Prince's Bay contended with a three-foot bluff at the end of their street. Grading streets after home construction also caused flooding owing to the differing heights with previously installed streets. This situation was exacerbated where sanitary sewers had not yet been established.

Dongan Hills continued to have perilous street problems. The city allowed one builder to construct 523 homes several feet below the legally mapped street grade level. When the city began installing storm sewers for South Beach Psychiatric Center and a nearby school in 1966, residents realized the "new roadbeds were high embankments built at legal grades, in contrast to the homes. Houses along [their] street were partly buried."[130] In fact, this condition already existed on Buel Avenue, "where residents' picture-windows look out on a street reminiscent of the Manhattan's old Third Avenue El."[131]

In a 1975 interview, a past employee of the Office of Staten Island Development stated that approximately two thousand homes in Midland Beach were built below grade. This, in turn, led to severe flooding conditions.[132] Flooding was a constant problem at Guyon Avenue and Hylan Boulevard in Oakwood by mid-1967. It would not be rectified until a health emergency was declared. The intersection was slow to drain during storms owing to "filling operations on private property with or without the consent of any city agency."[133] Drainage had been closed off by construction activity at the foot of nearby Adelaide Avenue. The sewer maintenance division of the borough had attempted to clear a ditch that needed to be eight hundred feet long. It had served as an outlet for storm sewers. But workers were prevented from doing their job as the clearing needed to be done on private property. An underlying matter was how drainage was handled at building sites with natural waterways as a creek in the area had also been blocked. Some believed water should be directed into a man-made culvert; others called for open streams to remain in a natural state to accommodate water. When Councilman Edward Curry viewed the property, he wondered why building was occurring on a lake he skated on as a youth.[134]

A near tragedy occurred on Tysens Lane because the ditch was blocked. Joseph Quinn of Tysens Lane fell into a two-foot water current after he removed a manhole cover to drain a catch basin in August 1967. As he was being pulled toward the open sewer, he regained his footing and was able to retreat to the roadside. Without the cover, the water quickly drained from Tysens Lane into the sewer, but the cover could not remain off.[135]

Finally, on September 15, 1967, a bucket crane was scheduled to begin excavating the ditch and clean out the creek that received the water from the ditch. The work was expected to take fifteen days. When finished, it was believed that the flooding problem would be rectified.[136]

Residents of Willowbrook were dissatisfied after construction workers partially diverted a brook that crossed Caswell Avenue. Trucks further flattened the streambed, forcing almost two feet of water to accumulate after a rainstorm.[137] The Department of Public Works was able to open part of the stream, but they could not clear it completely because the brook flowed through private property. Residents petitioned the Health Department to declare a health hazard so that a public works team could enter the property and fix the problem.

Accidental diversions, shortcuts, disregard, frugality, and other excuses led to serious problems for residents during the entire building boom. In early 1964, the City Department of Investigation discovered that a construction firm had buried a city sewer under tons of fill. The work had also commenced without a permit. Labau, Logan, and Oswego Avenues in Sunnyside were affected. According to the Highway Department, the "fill [was] piled several feet high on the unimproved streets." Furthermore, the department could do nothing because "although recorded on a city map [the streets] were never acquired by the city." At this time, the filled roads were blocked to vehicles. Public works officials also claimed they could not remedy the situation since the dumped fill had not crushed the existing sewer. The local civic association charged that the road was now as high as twenty feet in some places and that a natural drainage ditch had been blocked by the fill. It was further revealed that heavy winter rains caused "lakes of mud."[138] The mud had, unfortunately, inundated tiny Saint Nicholas Chapel at Labau Avenue and Northern Boulevard.

The City Planning Commission announced in the fall of 1966 that the required street width of sixty feet would be relaxed for Staten Island streets. It was hoped that this action would speed up the process of establishing the unmapped streets while also reducing "damage costs and inconvenience to abutting property owners" who were being affected by street widening.[139]

The following year, Mayor Lindsay called for builders to install new streets at elevation grades up to three feet lower than the legal grade in the low-lying areas of South Beach, Ocean Breeze, and Dongan Hills. Not raising the streets to legal grade was intended to alleviate flooding in these neighborhoods, which was exacerbated by the differing grades of older and newer streets. The lower legal grade would also support the construction of storm sewers.[140]

Flooding was a historic problem. Beachfront locations, structures built on clay, underground streams, and other natural and unnatural features contributed to the conundrum. While 1961 through 1965 were drought years for the island, 1966 kicked off with a storm that caught everyone's attention. In fact, New York Harbor was whipped into such a frenzy that ferries had difficulty docking at both St. George and Manhattan on January 23. High winds and high tide caused a surging tide four feet above normal. Streets and homes between South Beach and Great Kills were awash in water. Police were called in to rescue forty-one residents from homes in New Dorp Beach, Oakwood, and Great Kills.[141] A sluice gate blocked with driftwood prevented much of the Oakwood floodwater from receding. In Princes' Bay, five feet of water blocked streets and homes when Lemon Creek overflowed. Both the civil defense and the Red Cross were activated. Power failures in several neighborhoods not only led to a lack of lights, but also, in many cases, heat was knocked out.[142]

One of the most tragic episodes occurred during a severe thunderstorm on August 5, 1967. A fifty-six-year-old woman drowned in three feet of floodwater outside her Great Kills home.[143] In fact, the woman had spent four years dealing with the problem of flooding at her residence. Two days before her death, city officials notified her lawyers that "substandard paving installed by a private developer" was holding up her case. An investigation into her death led the Department of Public Works to reveal that an illegal storm sewer connection had been made by a private developer near the corner of Lamoka and Brookfield Avenues. Investigators were working to determine if the presence of the sewer caused the woman's death.[144] Not only did the women lose her life, but her dog also perished as she was holding the animal while attempting to escape the flood. A grand jury was called in to investigate.

On December 20, 1967, the grand jury "failed to take any action in the street ponding death" of the woman.[145] The previously mentioned substandard paving was not cited. It was also stated that the illegal sewer pipe played no role in her demise. The jurors believed that even though her death was caused by drowning, they could not establish a cause of death beyond a reasonable doubt since the drowning could have been from natural causes, lightning, or a combination of these factors. This was a blow to nearby residents who constantly battled the ponding problems around their homes. With the case dismissed, there would be no further investigation into the floodwaters that tortured them. One neighbor stated that during a heavy rain, the drowned woman's home resembled Hoffman Island. Both her and her neighbor's houses were older and had been built on lower grades than recently constructed houses in the area. The neighbor stated she was "thoroughly disgusted" with city agencies that did nothing about the water problems. She "acidly" added, "We've been pushed from pillar to post."[146]

The men of Windemere Avenue in Oakwood decided to act as a cohesive group whenever it rained. Six men would grab their shovels and picks to dig out a ditch that ran from one of their properties across North Railroad Avenue. In this way, they were able to siphon off the accumulated water that plagued the properties on the north side of the street. They were aided by natural seepage and an electric pump that operated constantly during rainstorms. It seems that the ditch continually filled with dirt and debris between storms. The rainwater in one homeowner's yard actually rose to six feet. It would then spread to five other yards and cover a portion of the street. At this point, it would begin coursing into cellar windows. One neighbor noted that thirty-seven children under the age of eight lived on the street. Neighbors were concerned that an accident was waiting to happen; thus the men decided to act. Homeowners acknowledged that the builder had installed dry wells for the water, but they did not solve the dilemma. Storm sewers were needed.[147]

Floods, drainage ditches, storm sewers, sanitary sewers, unnatural ponding, fear of children drowning, the drowning of the fifty-six-year-old resident and her dog, septic tanks overflowing, blocked sewer outlets, hillside runoff, mudslides, and storm sewers overwhelmed by downpours causing sanitary sewers to back up into homes were common headlines in the *Advance* during the 1960s. Who was to blame? Builders were to blame, planners were to blame, the natural conditions of Staten Island were to blame, unusually wet years from the mid-1960s through the close of the decade were to blame. The blame game goes on today. What could have alleviated these water problems was planning and studying the geology and topography of the areas before homes and streets were developed. Maps showing natural conditions and individualized plans to accommodate homes and developments would have prevented some of these issues. But it was, literally, a free-for-all on Staten Island.

Who could buy the land first? Who could put up the houses first? Who could build these houses as quickly as possible? Who could sell them first? Who could get out of Brooklyn first? It was a race like no other, and for the most part, no one was the winner except those who made a careless and quick profit. Since drainage was never considered, it would continue to be a headache for all concerned through the 1970s and beyond.

Building sites in the 1960s and 1970s were not governed by the same rules and laws that are now in effect. Site owners were not required to install safeguards at construction sites. In many locations, building was delayed or abandoned, allowing for dangerous situations. Neither plywood walls nor safety fencing were required, so kids oftentimes played in these hazardous locales. One unrestricted site at 30 Brookfield Avenue in Great Kills greatly concerned parents. In September 1966, a duplex was abandoned mid-construction and became an unofficial playground. Neighbors often saw boys walking on the cement divider walls of the open foundation. On either side of the wall, four feet of water stood, creating a potentially deadly site for drowning. If the victim was alone, they would have no way out of the water. In addition, opportunistic mosquitoes were taking advantage of the open water to lay their eggs. Soon, swarming bugs were biting all. The Building Department responded and designated the foundation a "hazardous condition." A violation was sent to the owner. He was given until June 22 to pump the water out of the foundation. If the water was not removed, a summons would be issued. The Health Department planned to notify its pest control unit about the mosquitoes.[148] Mosquitoes were the least of the public's worries. Not only did heavy rainfall during the summer of 1967 upset residents as they watched their basements fill up with water, but Staten Island parents were also concerned about the number of basements and excavated septic pits that would simply fill up with water at a time when construction sites were not closed off with fencing or boards.

Parents who lived in the area of Baltimore Avenue and Lindenwood Road in Great Kills were horrified that an open excavation was often filled with as much as eight feet of rainwater. Youngsters could easily fall into the pit and drown. Oftentimes, they played on the mound of dirt that had been removed and placed next to the pit. They were also known to place a wooden plank across the water-filled septic opening and walk across it. The parents of Staten Island had good reason to be concerned—and in early 1966, their fears were finally realized.

Two children, ages eight and five, drowned on January 10, 1966. They fell through ice that had formed on a man-made pond resulting from construction work in Castleton Corners. When the children went through the ice, workers were building houses nearby. While attempting to rescue the youngsters, four construction workers and a policeman became trapped in the pond. They needed assistance to escape the ice-covered water. Five-year-old Susan Larsen and eight-year-old Rita Reynolds broke through the ice and perilously fell into more than seven feet of water. The youngsters were trapped under the ice and were not seen for twenty minutes. When they were removed, attempts to revive them were futile.[149]

The office of Richmond County District Attorney Braisted and the New York City Police Department began an investigation to determine if any criminal negligence occurred. Public Works Department maps showed a swamp and a natural watercourse in the area. The map was based on a fifty-year-old survey that was apparently done in 1915. The day after the girls drowned, the *Advance* reported that drainage pumps were removing the water.[150] As a result of the investigation, a grand jury was called to determine if charges needed to be filed.

The investigation showed that the pond was "created by mounds of soil which may have blocked or diverted a natural stream, watercourse or natural drainage."[151] The soil piles were the result of installing streets for the development. District Attorney Braisted "referred to the possibility that builders might consider it an unnecessary expense to install temporary culverts or take other steps to insure proper drainage."[152]

According to the *Advance*, the builder had previously been fined $250 for constructing "nine houses without building permits and [it had] about fifteen similar violations pending."[153] Houses built by the company on Mountainview Avenue and Gansevoort Boulevard had drainage problems, so they were denied certificates of occupancy. Two of the houses on Mountainview Avenue were built without a permit.[154] It was later said that the "developer diverted a natural watercourse and raised the height of Gower Street to eleven feet above the existing elevation to create a water dam. No attempt was made to drain the water."[155] Specifically, the construction area was bounded by Queen, Portage, and Fanning Streets as well as Gansevoort Boulevard. The site was approximately twenty acres, and the builder planned on constructing one hundred or so houses.[156]

As a result, two men associated with the construction company were indicted and charged with causing the drownings. The indictment stated that both were responsible for the frozen pond that was created.[157] The men pleaded not guilty to two counts of manslaughter on April 15, 1966. Attorneys for the individuals were expected to request a dismissal of the indictment owing to insufficient evidence.[158] The indictment against one man was dismissed.[159] What happened to the other indictment is unknown.

In response to the tragedies, legislation was introduced to the city council to prevent such a tragedy from occurring again. The legislation included giving the Highway Department "authority to act in matters involving natural watercourses, streams and drainage systems that are responsible for hazardous ponding conditions," especially on private property.[160] The law was passed. It became effective on March 22, 1969.

The 1960s saw many empty buildings across Staten Island: houses that were being built, houses that were being sold, houses that were being moved. These deserted structures were prime locations for mishaps, chaos, and misfortune. With youngsters playing and teenagers drinking in these dwellings, anything could happen. Rats, garbage fires, and basements teeming with water all posed potential dilemmas. Personal problems, city red tape, legal snares, death, and so much more could halt the progress of construction. If advancement on a domicile slowed or ceased, the Buildings Department was usually called and the premises was boarded up.

Chapter 11

THE BANE OF MANY

SEPTIC TANKS AND SEWERS (OR LACK THEREOF)

Before the Verrazzano-Narrows Bridge opened and the landscape became carpeted with houses and developments, rules regulating cesspools were not stringent since open space abounded. For this reason, when septic tanks were installed, they were allowed to spill off into fields and other unused areas with little regard. As the island's population increased, the situation changed drastically.

In the late 1950s, experts on the subject of human waste removal warned that pollution from the few Staten Island sewers that existed was only going to get worse in the upcoming years. Even Robert Moses stated that pollution control authorities needed to recognize this fact and act, otherwise island beaches would become unfit for use. The existing sewage plants of the time functioned well when it was dry, but in times of prolonged rainfall, sewage and rainwater combined at the plants to cause an overabundance of water. The plants were not designed to process vast quantities of water and more. For this reason, regulatory chambers automatically shut down so the plants did not flood. When this occurred, untreated sewage water went directly into the waterways that surrounded the island.

As the bridge neared completion in 1963, the City Department of Health reported that only 35 percent of the borough was served by sewers. The following year, the Public Works Department requested $581,650,000 to construct the necessary borough sewers. Not one nickel of this was believed to be forthcoming. By mid-1965, 70 percent of all recently built homes in the borough were attached to septic tanks.

Builders were incensed when, in 1965, the city ordered them to stop using city water for mandatory absorption rate tests of septic systems. Approximately two thousand gallons of water was needed for each test, and the city was in the midst of a terrible drought. Builders siphoned water for the test from fire hydrants. They claimed building activity would be further slowed down by this regulation. Developers responded by bringing a lawsuit against the Water Register Bureau. The developers lost. The lawsuit was denied when a judge ruled that builders could not use hydrant water for testing. The ban on using municipal water was said to have halted the construction of thirty or so homes per week, and it was expected to "result in a 60 percent reduction in home construction." Before the ban went into effect, "there was an average of five seepage pit tests daily on the Island, with each test consuming about 2,000 gallons of water."[161] The lack of sewers, the proliferation of septic tanks that either failed or had little area for runoff, and the presence of underlying clay further exacerbated drainage. Builders were eventually required to install dry wells in poor drainage areas where storm sewers were nonexistent. The dry well, a concrete receptacle, was to absorb rainwater from roof drains. To work correctly, these wells were to be installed in sand and gravel. Because clay sat atop rock in large sections of the island, once a dry well was full, it was useless since the water could not flow downward. Adequate sewers would alleviate the problem, but no one expected them for another forty years. In some areas, they would not arrive for fifty years. In some areas, they have not yet arrived.

The combination of heavy rains, overdevelopment, and the lack of sewers caused serious floods in 1967. Prolonged rain enabled sewage, wetland water, and rain to rise from creeks and streams. Much of it flowed into the basements of homeowners. Wetland depth had not been measured since around 1912. It was easier to determine the depth of the New York Harbor than that of Clove Lake in 1964. In addition, many wetlands were built over before the State of New York passed a wetland law in 1975. The fact that so many wetlands were filled in prevented the public from knowing whether they were purchasing a structure built over a former swamp, marsh, or pond—which would prevent them from ever being able to stop the onslaught of water into their cellar. Even if money was available for proper storm sewers throughout the borough, many officials believed that at least four years were needed for sewer planning to catch up with house building. There was no hope that a complete storm drainage system would be planned or installed in the foreseeable future.

Real estate interests and builders were not pleased when the Department of Health implemented modern regulations governing septic tanks, especially tanks located around marshy areas. This included many acres of vacant island land. Builders complained that the measures were too stringent. They believed it would be impossible to build houses in certain sections of the borough. Health officials countered that rates of hepatitis and bacillary dysentery on the island were higher than in other boroughs. Some said the figures were misleading because Willowbrook State School had high incidences of these infections. Even so, the latest rules were actually prompted by the many septic tank failures of newly constructed homes—especially after a heavy rain. The Department of Health also required that builders post a bond guaranteeing that septic tanks would not fail within five years. Director of Development Holt Meyer was appalled at the proliferation of septic tanks and stated that he knew of no other municipality where tanks were allowed on such small lots. The tanks, he said, were a "black eye on the building industry."[162]

By 1968, 52.6 percent of existing Staten Island streets had sanitary sewers and 11.6 percent had storm sewers. As such, the island needed an additional 300 miles of sanitary sewers and 575 miles of additional storm sewers for the existing paved streets.[163] According to the *Advance*, "every sewer between West Brighton and St. George and Fort Wadsworth, extending inland to

Willowbrook State School was blamed for the high rates of hepatitis on Staten Island, as were septic tanks. *New York Public Library.*

Silver Lake, Grymes Hill, and Concord, empties directly in the Kills and Narrows without being treated."[164]

It was proposed that developers build their own community treatment plants since the city was unable to construct the necessary sewer lines and treatment plants to meet demand. A minimum of fifty homes was needed to secure permission for a private treatment plant. Builders did not believe fifty houses would cover the expense of such a facility.

It was not until January 1, 1970, that new septic tank installations were finally banned in New York City. There were exceptions. Tanks were allowed on lots of ten thousand square feet, or one hundred by one hundred feet.[165] In addition, if the landowner had filed plans to subdivide the property prior to August 1968, tanks could be used in below-minimum-size lots.[166]

The 1970 law did effectively ban construction on 90 percent of island plots. Builders were horrified. They countered that city sewer construction was at a standstill. Furthermore, they believed land values on the south shore would fall while values on the North Shore would skyrocket since sewers existed in that area. Builders claimed that 60 percent of home construction business on the island was now lost. Thousands of construction workers would be unemployed. Some builders believed it was a plot by the city to take over South Shore properties so they could be developed according to the ideals of modern planners. The Department of Health stressed that septic tanks were for rural areas or suburban lots of one to two acres, not urban properties. In addition, one official wisely noted that the lifespan of a septic system was limited. Owing to the exceptions, roughly one hundred tanks were installed in 1971, compared to about one thousand the previous year. This ruling did not help homeowners with existing septic tanks. Many piped their washing machine water directly into the street to keep their cesspool from being overwhelmed. Some homeowners had their tanks pumped every two weeks. Others would not flush paper down the toilet—instead, it was put out with the household garbage.

The first private sewage treatment plant for Staten Island was approved by the City Planning Commission in 1966. Richmond Klondike Corporation was building a two-thousand-house development on 136 acres in New Springville. The development was bounded by Richmond Hill Road at Steinway Avenue and Shiloh Street. The company agreed to post a $250,000 bond to ensure maintenance and operation of the plant for ten years. Homeowners would pay no more than $100 per year to use the facility. This and subsequent private sewage treatment plants were to be taken over by the city once sewage disposal infrastructure was in place. Interestingly, the builder of this development had the right to install septic tanks.[167] Five months later, the City Planning Commission approved the second private sewage treatment plant on the island. The Arthur Kill Land Corporation owned sixty-eight acres bounded by Arthur Kill Road, Armstrong Avenue, East Brandeis Avenue, and Cortelyou Avenue in Great Kills. The plant would have a capacity of 375,000 gallons per day and would service a 750-home development.

Chapter 12

THE POPULATION EXPLODES

As movement on erecting a bridge across the Narrows became serious in the mid-1950s, the population began to rise in small increments. In 1950, it was determined that 191,555 individuals resided on the island. By 1960, the population stood at 221,991.[168] Borough President Maniscalco believed the population would reach 500,000 by 1974. It did not, but it did feel as if half a million people had moved to the borough. The borough was predominantly white. According to one population expert, the white population was reinforced by the opening of the Verrazzano-Narrows Bridge.[169] Simply put, the bridge brought more Italian Americans, Irish Americans, Polish Americans, and so on. Overall, in the 1950s and 1960s, African Americans left the South and Puerto Ricans left Puerto Rico to relocate to the five boroughs in hopes of improving their economic status. The arrival of thousands of African Americans from the southern United States to Brooklyn caused what some historians call "white flight," described as white residents leaving the established white neighborhoods of the urban boroughs. In Staten Island's case, some white residents from Brooklyn relocated to the island to escape a perceived problem: that Black people were going to destroy their Brooklyn neighborhoods. Some of this fear was fueled by realtors who were trying to "bust" neighborhoods. In other words, these unscrupulous realtors hoped to instill anxiety in white residents so that they would sell to realtors, who would then resell to African Americans at inflated prices. Some suggest it was not the bridge that prompted the relocation of more white people to Staten Island; instead, it was the fear that the changing racial and ethnic makeup of their Brooklyn neighborhoods would lead to high crime rates and a decline in property values. On the other hand, many Brooklyn whites relocated to Staten Island for suburban living, which was no longer available in the other four boroughs.

African Americans had lived on Staten Island since the days of Dutch administration in the seventeenth century. In fact, enslaved Africans and African Americans were used as forced labor until the end of enslavement in New York State in 1827. Sandy Ground is the oldest African American community on Staten Island. It was established by free Black farmers in the early 1800s. They were able to procure land—an investment that was, more often than not, prohibited by white landowners. The Sandy Ground residents were joined by oystermen and their families who relocated from Maryland, a state that still practiced enslavement in the late 1840s. The free oystermen departed Maryland after laws were enacted to force their exodus. The government of Maryland did not want free African Americans living in the state, fearing that their presence would advance the idea of a revolt among the enslaved. The community of Sandy Ground flourished on Staten Island through the 1980s. Overdevelopment, the construction of the West Shore Expressway in the early to mid-1970s, and soaring land prices caused many residents to relocate from that community. Other African American communities existed on Staten Island at Manor Heights, Tottenville, Stapleton, and Mariners Harbor.

James Farmer of the Congress of Racial Equality spoke at Prall Junior High School, West New Brighton, on June 2, 1964. *Library of Congress.*

The influx of African Americans from Brooklyn increased the fears of many existing white island residents, who worried that their presence would destroy housing values. The summer of 1965 saw an investigation by the State of New York to determine which real estate agents were "discriminating in housing against minority groups, in violation of the Metcalf-Baker Law."

James Farmer, national director of the Congress of Racial Equality (CORE), spoke at Prall Junior High School on June 2, 1964. During his well-received speech, Farmer stated that bigotry and segregation could end with this generation if everyone stood up for its termination. He said that "decent people" who were "trying so hard to stay neutral" needed to be involved.[170] He believed it was essential that they not back away and move when an African American family moved into the neighborhood. In fact, he stated that this did more harm than the "frenzied, screaming idiots of Alabama."[171] Farmar spoke about the increase of housing segregation in the north and the income gap between white and Black Americans. He observed that African Americans were losing jobs in the private sector, and he highlighted the difficulties that young Black men faced when trying to get employment. The presentation was part of a celebration of CORE's third anniversary on Staten Island. Close to $1,000 was secured in pledges and contributions to the organization.

One of the organizers of the presentation was Richard Prideaux, a man who devoted his life to fighting discrimination and segregation through peaceful activism. According to his daughter Barbara, "He was never angry. He never yelled or cursed. He was a well-balanced man who was motivated to make change by curbing racism."[172] In 1961, Prideaux organized the Staten Island chapter of CORE, a group that emphasized direct action and nonviolence. Prideaux also served as chairman of the borough organization. He received numerous pieces of hate mail and telephone calls for his efforts.

In the mid-1960s, more than three hundred CORE members resided in Staten Island. Of these, 60 percent were Black and 40 percent were white. A number of clergy members joined, including the Reverend William Epps of Saint Phillips Baptist Church, the Reverend William H. Latham of the Reformed Church of Staten Island, and Rabbi Marcus Kramer of Temple Israel Reform Congregation. Many CORE members from the island participated in the Lincoln Memorial Freedom March in Washington, D.C., on August 28, 1963.

The initial meeting of the Staten Island branch of CORE took place at the Unitarian Church on July 14, 1961.[173] Representatives of both the Republican and Democratic Parties made statements. A Brooklyn resident, Abraham Bassford IV, talked about his motivation for fighting against segregation. Bassford had previously picketed a Port Richmond store with other college students and members of the Staten Island branch of the National Association for the Advancement of Colored People.

As the construction field thrived before, during, and after the completion of the bridge, discriminatory practices in the building industry were also flourishing. The morning of July 22, 1963, saw approximately twenty-four marchers protesting discrimination in the construction field at the building site of PS 25 in West New Brighton. Borough clergymen who participated in the demonstration included the Reverend Mills Skinner, the Reverend

William Epps Jr., the Reverend Dr. Winfield Burggraaff, Rabbi Marcus Kramer, and the Reverend Greg Carter. Protestors planned on picketing until an agreement was reached for "open hiring and the opening of apprenticeships" for African Americans.[174] Organized by CORE, the group returned the following day to protest construction projects involving government funds.[175] That day, protestors blocked trucks from entering the worksite. Meanwhile, at a Curtis High School construction site, protestors were taken into custody. It was July 24, 1963. This was the first arrest on Staten Island, as activists sought to get the message out that minorities needed employment opportunities in the field of building. Two men and three women were booked on disorderly conduct charges. They were transported to Stapleton Criminal Court, where the judge asked the women if they were mothers. "When each replied affirmatively," he asked if they were "proud to appear in Criminal Court."[176]

On July 23, a group that included Edith Overton and Catherine Johnson of CORE, along with Ernest Dow, vice president of the Staten Island Urban League Guild, had a meeting with store owners from the Forest Avenue Shopping Center (also known as the Plaza) at Port Richmond Center. Plaza merchants, who were described as "very cooperative," agreed that a "fair employment practices policy and the practice of mentioning 'open hiring' in employment advertisements" would begin.[177]

On July 26, picketers took their protest to Borough Hall. The group decried the city's lack of response to job discrimination for projects funded with public monies and hoped to meet with Borough President Albert Maniscalco to get his support. Thirty protestors were present along with a dozen policemen. The message was being heard. After the protest at Curtis High School, five African Americans were hired to work at the site.[178] At Borough Hall, Maniscalco agreed to assist with the group's efforts after meeting with six protest leaders. They requested that he help with the fight against prejudice, especially with employment discrimination in the construction industry—the number one industry on the island. They asked that Maniscalco remind Mayor Wagner that the "city has a non-discrimination clause in all of its contracts"—a clause that the group believed had never been enforced.[179] Maniscalco agreed to this request.

Discrimination in the sale of houses was another fact before, during, and after the opening of the Verrazzano-Narrows Bridge. It wasn't until 1968 that the Fair Housing Act became federal law. This act was created to "protect people from discrimination when renting or buying a home, getting a mortgage or seeking housing assistance."[180] Before this act was passed, CORE would send two couples to a Staten Island house that was being sold. They would inquire about the house separately. One couple was white; the other was Black. CORE was testing if the premises were equally available to African Americans. On most occasions, the Black couple was told that the house had been sold and was no longer available, but the white couple was offered an opportunity to purchase the home. CORE members would then usually protest in front of the house. This type of discrimination prevailed in the 1960s. In fact, two local realty offices were targeted by CORE during the fall of 1963. In one instance, an African American family was told they could not negotiate for a certain house because a binder had been placed on the property. The family informed CORE. CORE sent a white member to the realtor, who stated that the house was "readily available."[181] When brought to the house, the white member received a standard sales pitch. The African American family filed an affidavit with the State Commission for Human Rights against the realtor, while CORE picketed the realtor's office.

Chapter 13

LET'S DO THE PORTABLE SHUFFLE

The population of overburdened Staten Island schools was soaring with the arrival of baby boomers. In addition, many borough schools were antiquated and undersized. Around 1955, the New York City Board of Education began developing a plan for a system of junior high schools throughout the city. They believed these schools would take the burden off overpopulated elementary schools. The first such school to open was at the corner of Willowbrook Road and College Avenue. Junior High School 51, also called the Edwin Markham School, was designed to accommodate students from seven neighboring elementary schools: Public Schools 21 in Elm Park, 22 in Graniteville, 26 in Travis, 30 in Westerleigh, 19 in West New Brighton, 20 in Port Richmond, and 44 in Mariners Harbor. Long before the bridge opened, PS 44 was on a triple session day. It had the largest student body of any island grade school. Junior High School 51 was designed to accommodate 1,500 students. Four more Junior High Schools were in the planning stages, including Junior High School 49 in Stapleton, which was dedicated to Berta Dreyfus on January 29, 1963. Junior High School 27 in West New Brighton was dedicated to Anning S. Prall on October 21, 1963. Junior High School 2 in Midland Beach, named in honor of George L. Egbert, was dedicated on May 11, 1964. In late 1964, Albert Maniscalco urged the city to speed up the opening of Junior High School 24 in Great Kills. It would accommodate 1,800 students. Eventually, it was dedicated to Myra Barnes.

Junior High School 7 in Huguenot was scheduled for dedication to attorney Elias Bernstein on April 19, 1966. During its construction, youngsters and parents alike were concerned about safety at the building. The school was built on what locals called the Old Mud Hole. For this reason, many believed that the school was sinking into the mud. Some students refused to enter the structure. They stated, "It's in the middle of water." They were not far off the mark: A structural engineer examined the site in January 1964 and discovered that "earth movements on the sides of the deep excavation [had] shifted the pilings. To combat this situation steel re-enforcements were used." The Board of Education claimed the problem was rectified but "loose footings and drainage problems" both plagued construction.[182]

High schools were already overtaxed as the 1960s moved along. Soon, more baby boomers and Brooklyn transplants would be arriving. The local school board and parent teacher associations called for additional high schools.

New Dorp High School and Curtis High School both had new additions by the spring of 1964, but they would not solve the dilemma. Fortunately, a proposed three-thousand-seat high school at Sea View was in the planning stages. A modern South Shore high school was also being discussed. Parents also hoped to replace the Ralph R. McKee High School. The vocational school was outdated. When the discussion of a high school at Sea View began, it was hoped that this school would be both vocational and academic in order to replace McKee. The idea was rejected.

In October 1965, the new South Shore high school on a twenty-one-acre tract in Huguenot was approved. It would seat four thousand pupils. Bounded by Luten Avenue, Eylandt Street, and Amboy Road, this was the future Tottenville High School. When completed, it would be the largest high school on Staten Island.

Paperwork on the creation of a new high school at the former military site of Miller Field in New Dorp was moving along in early 1965. It would be seventeen years before this school was officially opened on New Dorp Lane. Original plans called for this school to be vocational and technical, but this did not happen. It was noted at the time that the city was planning to "do away with vocational and technical high schools in favor of four-year comprehensive high schools."[183]

McKee High School was never phased out. One newspaper article noted that the school's role was more important than ever in 1965. Geared toward teaching students a trade—many of which were desperate for new professionals—McKee offered three types of diplomas: technical, commercial, and vocational. Programs in electrical installation, machine shop, automotive, printing, beauty culture, machine design, plumbing, woodworking, sheet metal, and dressmaking were offered. In 1965, administrators were hoping for additional space as the Saint George building was never meant to function as a high school. It was small and cramped, lacked ball fields, and had inadequate athletic facilities. Interestingly, city schools like McKee were developed in the 1920s to accommodate students who needed to leave the classroom for employment to help their families.

Portables—impersonal metal classrooms, where children felt like icicles in winter and barbecued chickens in summer—were moved as easily as lawn chairs around a picnic table during the 1960s. Their "beauty" (according to administrators) lay in the fact that they were not stationary. They were moved from school to school and borough to borough.

In May 1963 came the announcement that portables would be installed at Public School 42 in Eltingville and at Public School 8 in Great Kills. They arrived in Eltingville that summer. Overcrowding was causing split sessions. In addition to the six portables, three classrooms and a teacher's service room were being established at a nearby church to end the split sessions. It

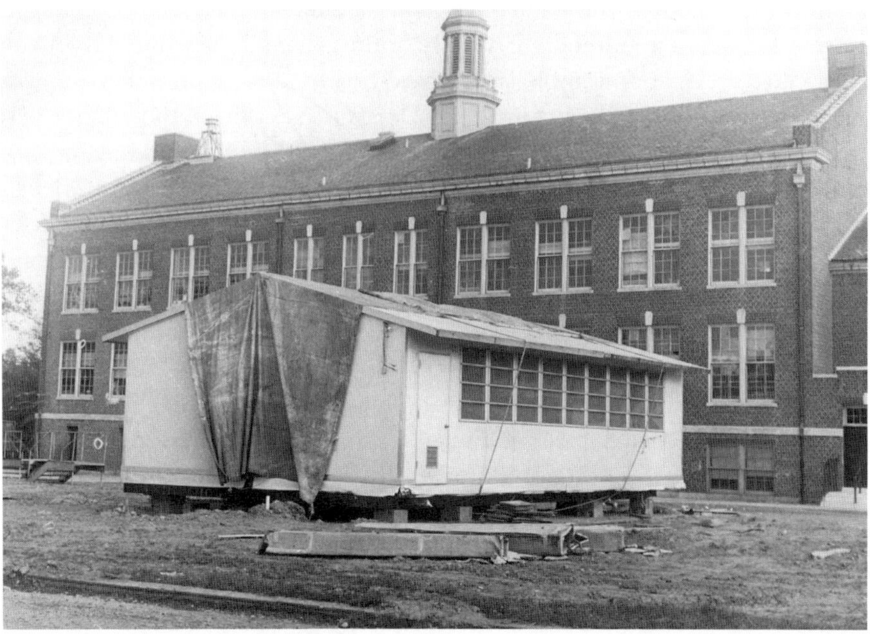

A portable is installed at PS 42 in Eltingville, circa 1963–64. *From the* Staten Island Advance.

was feared that the church annex might not be ready for the beginning of school, when classrooms were expected to hold more than forty pupils. As the population of PS 42 climbed, the PTA feared that if the split sessions did end, it would be temporary as the number of schoolchildren continually increased. In fact, eighty-nine new houses had recently been completed nearby. Administrators were anticipating that the number of students would rise from 641 to 800 when school began in September. It was also expected that PS 8 in Great Kills, PS 5 in Huguenot, PS 28 in Richmond, and Port Richmond High School would all receive portables.[184]

It was soon realized that there was a shortage of portables. Island schools were in a dire situation. A survey of the other boroughs was undertaken to determine if any portables were available. Working "under an emergency order," the board reported that a newcomer to the prefabricated classroom field was moving slowly on completing portables. Administrators stated if the portables did not arrive, three elementary schools would go on split sessions.

Again, in August 1964, parents of students at PS 8 in Great Kills were awaiting ten portable units, while PS 42 was expecting the arrival of six more portables from Queens. Meanwhile, five units were on order for PS 11 in Dongan Hills, and two were scheduled to arrive at PS 38 in Midland Beach. As they were needed for the upcoming school year, parents were in advanced stages of concern about their arrival.[185] At this late date, all would have been happy to see even unassembled portables lying in the schoolyard. Two units promised to PS 38 were yet to arrive by December 1, 1964, although one was expected the next day and the other was scheduled to make an appearance on December 4. Students in the fifth and sixth grades at the Midland Beach school were on split sessions; some students were not released until four o'clock in the afternoon. At the close of the school year, the portables would no longer be needed since 150 students were being transferred to the new PS 52 in Dongan Hills. Fortunately, PS 52 opened seven months ahead of schedule, eleven months after construction began.[186] It was built in record time, as was a new PS 23 in Richmond. Administrators of PS 23 experienced a not-so-uncommon problem on Staten Island. The school was scheduled to open its doors to several hundred students on September 15, 1964, but roads and sidewalks had not been installed around the school by July 24, 1964. As such, the Highway Department was attempting to move quickly through the winding red tape of the City of New York to obtain over $110,000 for this construction. There were existing roadways, but they did not match city regulations for school sites. It was believed that the cause of the problem was the earlier-than-anticipated opening of the school.

August 31, 1964, saw a school protest march. This one involved PS 8 in Great Kills, and the protesters were not marching to a happy tune—nor was the protest at PS 8. It was at Borough Hall, where parents displayed displeasure that ten promised portables had not been installed for the new school year. One hundred parents picketed while PTA leaders met with Borough President Maniscalco to find a remedy for the situation. Maniscalco pleaded with the PTA president to call off the demonstrators until their meeting was completed, but the parents refused to depart. The contractor responsible for assembling the portables claimed he was waiting for panels from a Florida manufacturer. The rally had been organized the previous week when a Board of Education official informed parents at a PTA meeting that the portables would not be available until mid-October. It seems that the word *busing* was also presented during the discussion. When parents heard *busing*, all was lost, since busing to them involved integration. After the meeting, one attendee said, "We all went mad." Another sign proclaimed, "Portable classrooms, not portable children."[187] The portables were ordered a year before and were meant to offset the lack of space for a student enrollment of 1,700. Within one year, the population of this school was expected to escalate to as many as 2,200 students.

By September 14, the portables had still not arrived at PS 8, so irate parents boycotted the school. According to the principal, 724 of the 1,655 children registered attended class, but one PTA parent countered that by her count, only 250 students were in class that day.[188] The boycott was accompanied by a protest of approximately thirty mothers, one with a four-year-old in hand and a sign that read, "Will PS 53 be ready for me?" This sign reminded all that the parents had pleaded for the completion of PS 53 in Bay Terrace and its one thousand seats as it would alleviate overcrowding at PS 8. The boycott and march continued the following day with eighty mothers on the move outside of PS 8 while thirty-plus mothers were stationed half a mile away at Katan Avenue, where several previously installed portables were located.[189] The hoped-for portables did arrive—but by June 1, 1965. Even with fifteen portables, PS 8 was still overcrowded, so ten more were under construction.

Integration of New York City schools began in 1964. The Board of Education was attempting to integrate all five boroughs, but Staten Island was, for the most part, not involved in the effort with the exception of two schools. Both New Dorp High School and Tottenville High School were to receive African American and Puerto Rican students who attended "schools with a high percentage of ethnic imbalance." If such students wanted to transfer from their schools in their home boroughs to either of the two Staten Island schools, they were welcome to do so. At this time, the Board of Education stated that the schools on Staten Island were 90 percent white. No students signed up for the transfer.[190]

February 4, 1964, was a big day for the students of the former PS 18 in West New Brighton. They were led on a march from their 1890 school to a modern structure several hundred feet away. PS 18 was Staten Island's thirty-fourth elementary school.

In the early 1960s, some youngsters still attended classes in nineteenth-century wooden schoolhouses. Since they were fire hazards and obsolete, the city began leveling the ancient structures. One such casualty was PS 10 in Egbertville, built circa 1894. The Board of Education called for closing the school in June 1964. For some reason, the old school had been newly painted that spring. No one was positive that it would be demolished. There was even discussion of moving it to the Richmondtown Restoration. The first graders and kindergartners in residence at PS 10 in 1964 would be heading to the new PS 23 in September. Some parents felt regret, believing that the timeworn school had a country atmosphere. Built when Grover Cleveland was president, the structure had four upstairs rooms for a custodian. The first floor had served as a one-room classroom for fifty-nine years. In 1953, another room was added. Miss Anna Raichle was the teacher, principal, office clerk, song leader, lunchroom supervisor, and just about everything else until her retirement in 1961. Threatened with demolition in 1958, the school received a reprieve owing to Staten Island's expected population growth. PS 10 would handle the overflow of newly arriving students. It was not the last wooden schoolhouse in New York City. That designation went to PS 3 in Pleasant Plains. It, too, was built in 1894. PS 3's elimination card came up the following year.

Mothers continued to show their displeasure toward overcrowded schools in early 1965. They were disgusted that new schools were not being built fast enough. They believed that the Board of Education was underestimating the number of seats needed for the population explosion that was occurring. For this reason, they went directly into areas that were under development. These parents wanted direct answers from builders, contractors, and construction workers. They sought an approximate count of how many houses were going to be constructed on the island. Simply put, they wanted to know the number of families that would be resettling in the borough so they could compile the number of children who would need school seats. There was a history of disputes between the Staten Island Federation of

Parent Teacher Associations and the Board of Education. The PTAs believed that the board undercounted students. False numbers prevented accurate planning and scheduling for new schools in crowded communities. The numbers compiled by the PTAs led to a decision by the Board of Estimate and the city council's Finance Committee. These entities declared that the mothers of Staten Island were correct in their determination of the number of houses that would be erected or planned through 1968. Mayor Wagner quickly announced that he was seeking a site and money to build a new school in Castleton Corners. The women had concluded that the number of students in elementary and junior high schools would double by 1968. The school system would need another twenty thousand seats to match the projected enrollment. The federation insisted that another $12.3 million was required in the 1965 budget for eight more school projects.[191] The mothers of Staten Island were determined to stand up to the shirts that ran the City of New York—specifically those involved with the Board of Education.

Even so, the Board of Education claimed that North Shore schools had 2,375 vacant school seats. The Federation of PTAs vehemently denied this. In response, they asserted that the board counted library, sewing, and craft room seats as available classroom space. The federation also said that the count was done in October 1964. By late March 1965, these numbers were outdated. According to the board, PS 30, the most overcrowded school on Staten Island, had 106 empty seats. In spite of this, the school was to receive twelve portable classrooms in 1965. The fight continued.[192]

In 1966, the New York City Board of Education operated 886 school buildings in the five boroughs. Approximately 175 of these structures were built before World War I. At this time, the board was screening 159 of these buildings to determine which were in dire need of repair. No doubt they all needed major rehabilitation, but the board was desperate as baby boomers, the children of new immigrants, and families with youngsters arriving from the South all needed to be educated. The school superintendent explained that painting, adding toilets, and improving electrical work was to be done to update these buildings. The first seventy-five schools to be refurbished would be announced by July 1, 1966. The board hoped to have all the antiquated schools refurbished within six years. On Staten Island, there were fourteen schools built between World War I and 1938, while eleven schools were erected between 1878 and 1896. All were still in use, the oldest being PS 1 in Tottenville.[193] It was anticipated that PS 1 (built in 1878) would be selected. PS 4 in Charleston (which opened in 1896) and seventy-six-year-old PS 18 in West New Brighton were also on the wish list. No one was sure if these three structures would be included as a number of older schools on the island were scheduled for replacement.[194]

One long-standing school that survives today is PS 65 on Saint Paul's Avenue in Tompkinsville. Originally referred to as PS 15, it was built in 1897. In early 1965, the parents asked the Board of Education to leave their school alone. The board planned to have the PS 15 district absorbed by the PS 16 district. This meant that 271 students would be transferred to the soon-to-be-opened PS 16. The parents wanted no part of it, even though PS 15 lacked a cafeteria and a gymnasium. They believed their school had the "personal touch." They firmly stated that principal Helen Meade knew every student, while every teacher knew each student's family and any problems that the student might have. They called the school stable, not overcrowded. In fact, each of the ten school classes had fewer than thirty children each. Parents were also concerned that students would have to cross the always active Victory Boulevard in order to reach PS 16. The PTA president insisted that all were happy with the way things were.[195] Parents were soon collecting names for petitions to keep the school open while also consulting attorneys on the matter. The Board of Education informed parents that the school had been "condemned" because it was "outmoded." The building was viewed as "along in years."[196] Its future included demolition, but the date had not been set. From an engineering standpoint, the school had "exterior masonry walls with interior framing above the first floor consisting of wood or unprotected iron or steel." The Fire Department rated it A-1. This meant that the building met all requirements of the fire code. The Building Department determined it to be structurally sound. Parents and the committee they formed wanted the city to decide the school's fate based on its merits. The city voted to shut the school down at the end of June 1966. For some reason, it still serves as a public school in 2025. PS 15 is known as the Daniel D. Tompkins School. Tompkins was the man who developed the community known as Tompkinsville. The school lot has been the site of a public school since 1855, when the Middletown Township District School No. 1 stood on the site. A Queen Anne–style building, it is a designated New York City landmark. This status will hopefully prevent the building from ever being destroyed.

The oldest surviving wooden school in New York City at this time was located on Staten Island. It was built in 1894, and it stood on the same site as an earlier school built in 1874. It was the original PS 3, and it was earmarked for destruction in the summer of 1965. The last attending schoolchildren had departed, and the building was done serving its useful purpose. In 1904, an "addition" to the schoolhouse had been constructed. It overshadowed the 1894 structure. It, too, had a date with the wrecking ball. In 1960, both school buildings were joined by another addition. Removing the 1894 and 1904 structures would allow room for a $1.5 million wing. It would be home to ten classrooms and a dual gym and auditorium. Known as the Pleasant Plains School both now and in the past, the school is located on Goff Avenue.[197]

In early 1965, parents at PS 30 in Westerleigh were wishing they had a new school as their community and the surrounding communities that fed the school grew at a frantic pace. Officials and parents were, of course, forced to settle for portables. The Board of Education was offering to alter district lines so there would be fewer children eligible for school enrollment. Part of the alleviation process also involved transferring approximately 100 children who lived around Brielle and Bradley Avenues to PS 23 in Richmond. Administrators were also contemplating the reassignment of another 100 students to PS 29 in Castleton Corners. Back at PS 30, administrators were in a quandary as an estimated five hundred homes were to be built in the district during the next two and a half years. With 1,347 students registered, PS 30 already had one of the largest enrollments of any kindergarten-to-sixth-grade school on the island. A minimum of eight portables was requested for the school year beginning in September 1965. Since no space on or near the school grounds was available, the school superintendent was also attempting to find a location for the portables.

The previous December, Marie Cotter, president of the borough federation of PTAs, stressed the need for change at PS 30. She vigorously informed the Planning Commission that more money was needed to relieve overcrowding problems at the school. She noted with words and photographs that students were forced to eat lunch in the gym and in a former bicycle room of an obsolete 1894 addition. Youngsters sat on the cold concrete floor as no tables were available. Cotter further noted that twelve kindergarten classes had only one bathroom. She called the situation disgraceful. The PTA president was irate that money for a planned addition had been erased from the upcoming budget for no apparent reason.

Construction began on PS 52 at Mason and Buel Avenues in Dongan Hills in March 1964. The 950-seat school would have twenty-three classrooms, three kindergarten rooms, and a library in a three-floor brick building. It was hoped that the new school would relieve overcrowding at PS 11, where administrators were hoping that five portable classrooms would arrive by the fall of that year. An additional two hundred students were soon expected. At this time, three sixth grade classes from PS 11 were attending school at Egbert Junior High School, while two first grade classes were on an overlapping session in a makeshift classroom at the Berry Houses in Dongan Hills.

Citywide, 276 projects were proposed in 1964. They would take place during a six-year building plan. For school year 1965–66, PS 32 and Junior High School 24 in Great Kills would begin construction, as would additions to PS 3 in Pleasant Plains and PS 48 in Concord. Other long-term recommendations included an addition to PS 50, Oakwood; a new PS 56, Charleston; a replacement for PS 20, Port Richmond; and construction of PS 60, Bulls Head.[198] Not all projects proposed by the Board of Education were finalized or completed on Staten Island. That is not surprising, since there was not one Staten Island resident on the Board of Education.

The addition to PS 48 in Concord was not being built quickly enough, so many of the students' mothers planned a demonstration of more than one hundred participants for August 16, 1965. They had been waiting five years for the addition, and it was time to act. Moms, many of them pushing baby carriages, planned to carry signs "criticizing the Board of Education's 'poor planning' techniques."[199] Funding for the proposed construction had been recommended two years in a row, but it never reached final approval and was dropped from the 1966–67 school building program budget. The parents were to be appeased by the placement of four portables in a newly constructed school playground that had been installed a few months before. Parents were angry. Not only had the construction project been canceled, but now the outdoor play area was also becoming unavailable. Another bone of contention was that PS 48 lacked an appropriate cafeteria. It was so small that "pupils [were] accommodated in four separate feedings"; after eating their hot lunch, they were sent to sit in the school's gymnasium until classes resumed.[200] In 1964, the student population rose by two hundred. The PTA president expected that the number of kindergarten students alone would double to more than four hundred when school started in September 1965. It seems that during the previous nine months, 564 new houses were built in Concord. The rise in population was creating turmoil for the school. Meanwhile, a seventy-unit apartment building was being finished on Richmond Road. The additional students from these apartments would surely bring further chaos.

In September 1964, enrollment in the borough's elementary schools and junior high schools reached 27,351 students. The Board of Education had expected 25,792. Schools in Great Kills, Eltingville, and Richmond saw the greatest increase. Public high school enrollment stood at 9,510, making for a total of 36,891 public school students on Staten Island.

Enrollment at Catholic and private grammar and secondary schools had reached 21,600 in 1964. Combined, there were 58,491 students attending school on Staten Island in September 1964. These numbers would continue to rise. The Archdiocese of New York opened Countess Moore and Saint Joseph by the Sea High Schools as well as an elementary school in Oakwood. Additions had been added at parish schools in South Beach and West New Brighton by the fall of 1964.

According to the Staten Island Parent Teacher Association, the 1965 enrollment in public elementary and junior high schools would double within three years. They estimated that by 1968, there would be fifty-five thousand school children attending these local schools.

In 1965, the following projects were concluded or were underway:

- PS 52 in Dongan Hills opened for 830 pupils.
- Under construction were PS 16, Tompkinsville for 1,200 pupils; PS 31, New Brighton for 900 pupils; PS 53, Bay Terrace for 900 pupils; PS 55, Eltingville, for 900 pupils; and Junior High School 7, Huguenot for 1,500 pupils.
- Curtis High School was undergoing reconstruction for an additional 300 pupils.
- In the planning stages were an addition for 450 pupils at PS 42 in Eltingville; an addition at PS 50 in Oakwood for 320 pupils; PS 56 in Charleston for 600 pupils; Susan B. Wagner High School at Sea View for 3,000 pupils; and an addition at Port Richmond High School for 320.
- Projected for the capital budget were an addition at PS 3 in Pleasant Plains for 400 pupils; PS 32 in Great Kills for 1,200 pupils; and Junior High School 24 in Great Kills for 1,800 pupils.[201]

PS 56 would replace PS 4 on Arthur Kill Road in Charleston near Storer Avenue. The sturdy little school was composed of solid Kreischer bricks that were manufactured in the community when it was called Kreischerville in the nineteenth century. The historic landmark still stands and is now used by IS 25 students.

Teachers and students welcomed the arrival of exhaust fans for the stifling portables of Staten Island in June 1965. They were to be installed only in structures with windows on one side; those with cross ventilation were excluded. It was documented that 1,890 students were learning in fifty-four portables at six Staten Island schools as of June 1965. On ninety-degree days, some pupils suffered heat prostration, while others fainted. On many occasions, teachers had to move students outdoors for open-air instruction. This was not surprising, since inside temperatures were generally ten to fifteen degrees higher than outside. Students were sometimes forced to endure 105-degree temperatures. Staff and pupils called the buildings hot boxes, the Ovens, greenhouses, and incubators.

With the student population exploding, the Board of Education attempted to establish a "land bank" for future island school sites. Properties would be purchased before they disappeared and before real estate prices further escalated. Land on the South Shore was the main focus as the board hoped to purchase enough property for seven additional primary schools and five additional junior high schools. More than twelve thousand students would be accommodated. Some of the land to be set aside was already owned by the City of New York.

Unfortunately, the Planning Commission turned down the land bank idea in late 1965. The commission believed that the city owned ample land for school construction, so there was no need to accumulate further acreage owing to financial constraints. All were further dismayed when the commission pushed back construction of the South Shore High School and elementary schools at Charleston and Meiers Corners to balance the budget in 1965. Even so, a new junior high school in New Brighton was recommended.

Public School 17, the "red brick school on the hill," was built around 1898. It would soon be empty. PS 31, officially named the William T. Davis School, was scheduled to open in the fall of 1965. Located at Layton and Crescent Avenues in New Brighton, it was deemed a "more effective school." It would have smaller classes, extra teachers, and more student services. When the school was completed, a prekindergarten for four-year-olds would be available. A social worker and two psychologists would also be on staff. A bronze plaque of Mr. Davis was on view in the lobby so youngsters could learn about the island's most admired naturalist.

By 1966, Mid-Island schools were being inundated with new students as more and more housing opened at Bulls Head, Graniteville, New Springville, and Travis. Once farmland, these communities were now under assault by builders who were erecting hundreds of homes.

PS 26 in Travis was receiving six to seven new students each week by January 1966. Enrollment at this 1896 school had risen 25 percent since

September 1965. (This was a school where a rope was pulled each morning to ring the school's tower bell, a reminder for the neighborhood children to be on their way to learn the three *R*'s. Travis residents of 1966 were said to still be setting their clocks by the pull of the bell.) The school's rolls were increasing daily, yet the city turned down a request for a new school. As a matter of fact, the original PS 26 is still being used as of this writing.

PS 22 was groaning under the weight of new student arrivals when an *Advance* article in August 1966 announced that forty-eight new houses were being built on Rockville, Freedom, and Rivington Avenues in New Springville. Another thirty-three houses were also under construction on Dawson Circle, Carnegie Avenue, and Amsterdam Avenue in that community. Most of those homes would be two-family dwellings. While this was a miniscule amount of construction for a contemporary development, new houses were being built rapidly. In fact, New Springville would experience the greatest population growth of all Staten Island neighborhoods after the opening of the Verrazzano-Narrows Bridge. The bridge was the biggest contributor to the student population of PS 22. Even though builders were claiming financial hardship at this time, plans for another twenty-two one-family houses were submitted for Rockland, Carpenter, and Ashworth Avenues, as well as Glenn Road in New Springville. In addition, eighteen two-family homes on Rockville Avenue and Gary Place were also on the drawing board for this community. In 1966, PS 22 in Graniteville was to receive ten portables. Due to the expansion of school district lines and the population boom in New Springville, the student body of the Graniteville school was expected to expand beyond its capacity. (This author spent fifth grade in two of the portables. So many children transferred into my first fifth-grade class that it had to be broken into two classes, hence the move to another portable.)

Relying on portable classrooms brought numerous issues for administrators, teachers, and, of course, students. In the case of PS 30, four portables were going to be used beginning in the fall of 1967. It was a huge "headache" for principal Vincent Sarnese. Most of the problems revolved around the fact that the four units were placed on Muller Avenue—half a mile away from the main school on the Boulevard in Westerleigh. The first problem was that brothers and sisters would be separated if one was placed in a portable and their sibling was expected at the main building. Another situation involved students who were in the school orchestra or the glee club: They needed to take classes in the main building.

PTA president Marie Cotter and 300 parents once again descended on City Hall in a show of force for the budget hearing held on February 17, 1966.

Undersized PS 30 on Wardwell Avenue in Westerleigh was one of many Staten Island schools that relied on portables. *New York Public Library.*

They wanted the city to know that Staten Island was being shortchanged. This time, there was a new mayor, John Lindsay, a Republican whose efforts nonetheless displayed a liberal outlook. The mayor attempted to reduce the number of speakers from Staten Island at the meeting. In fact, the PTA president informed the mayor that their remarks would be brief and "not too boring" since previous City Hall meetings involving the Staten Island Federation of PTAs had run far beyond their scheduled conclusion. PTA members were not pleased with Lindsay's request, nor were they happy that he did not stay for their entire presentation. The group rallied around the fact that 1,500 new residents were descending on the island monthly. They stated that by 1969, there would be 16,300 high school students in the borough, and for this reason, the South Shore High School must be built to accommodate them. They also stressed the plight of the old Travis school, PS 26. They informed officials that the modest-sized school was being inundated with pupils from the thousands of newly built homes that were rising in surrounding neighborhoods.[202] Elementary schools at Meiers Corners, Concord, and Charleston were demanded, as were additions at three South Shore schools. According to Marie Cotter, PS 52 had opened five months before, and it was already nearing capacity. Only four blocks from the school, 150 duplexes were under construction. These structures would bring in hundreds of new students, as would the hundreds of other homes planned for the community.

It was projected that 40,644 students would be entering public schools for the 1966–67 school year.[203] None of the twenty-two portables ordered were expected to arrive in time for the first day of school. In total, both public and nonpublic schools would welcome 64,265 students on September 12, 1966.[204] Two schools, PS 53 on Bay Terrace and Saint Thomas the Apostle and Saint Joseph's in Pleasant Plains, were to open for the school year, although the Catholic school would open a week later owing to construction delays.

The Board of Education had a huge problem for opening day that September: There were twenty-two vacant regular teacher positions throughout the five boroughs. To alleviate the problem, school officials sent three individuals to take emergency exams in order to qualify for teaching licenses. On Staten Island, substitutes covered a number of classes. Borough high schools were desperate for math, music, and home economics teachers, while two high schools had openings for stenography teachers.[205] The increased student population caused the Pioneer Bus Company, the transportation system for island students, to purchase three additional buses before school began.

Special conditions existed on Staten Island when it came to busing. Many children remember the lengthy bus trips they had to take to arrive on time at local schools. Students who lived near the corner of Merrill Avenue and Richmond Avenue in Bulls Head were picked up an hour before classes began. The bus then traveled all the way to Signs Road in New Springville to pick up additional students. Then the bus driver turned around and went all the way back to PS 22 at Monsey Place in Graniteville. That's almost five miles round trip. Another student remembers when the school bus would pick her up at Merrill Avenue. The bus would go as far as where the Staten Island Mall now stands. From there, it would turn around and head back to PS 22.

The children who walked to school in a borough that was still rural in many locations faced issues of their own. Youngsters were forced to walk where no sidewalks existed, where septic tank runoff flowed, where some streets were not even paved, or where streets suddenly ended. In 1969, residents were tired of these and other conditions, which they rightly considered dangerous. As such, borough development director Holt Meyer requested that anything related to student transportation be switched to the district superintendent of Staten Island.[206] Part of his argument was that children who lived near schools should not always walk under the conditions previously mentioned just because they lived close to their schools. Staten Island was not like the other four boroughs, where streets, traffic lights, and sidewalks were neatly laid out. Meyer noted that someone familiar with local conditions should make decisions regarding pupil transportation. The Board of Education disagreed and denied Meyer's request.

The Board of Education had granted a temporary variance to more than forty Great Kills children who walked to PS 32 under hazardous conditions. Youngsters were not allowed to ride a school bus unless they lived half a mile or more than a mile from school, depending on their grade. A variance was difficult to obtain because of budget limitations. In 1969, the city paid $27 million to transport students by bus to their schools.

That fall, junior and senior high school students were crushed into Transit Authority buses even though sixty city buses had been added since 1963. (That led to a total of 250 buses in the island fleet.) Buses averaged seventy passengers per ride. But there was no law to limit the number of students per trip. Most students who rode city buses at this time remember the crush of humanity on these vehicles as they headed to and from school. What a relief it was to depart these buses—if you could get past the other passengers to reach an exit.

Increasing enrollment at island schools put pressure on the meager transportation hub at Saint George. Six public, parochial, and private schools were located there, as were 2,000 day and evening students from Staten Island Community College. Curtis and McKee High Schools alone had 3,160 students in the fall of 1964. Saint Peter's High School for girls and the parish grammar school accounted for another 1,000 students, while Staten Island Academy on Wall Street had 126 pupils. Public School 16 in Tompkinsville added another 618 individuals to the congestion.[207] Students from these and other schools also utilized the Saint George Public Library. At times, it was impossible to find a seat within the building owing to the crowds seeking information. By 1966, the principal of Curtis was calling for patrolmen at the school to supervise student safety at arrival and departure times. In addition, in 1966, five buildings were under construction by McKee High School, adding to the congestion. One commuter noted, "There are kids darting every which way. I try to avoid the area, if I can."[208]

To some students, an unnecessary evil was taking place during the summer of 1967. Summer school was mandatory for four thousand pupils. Public school students went to New Dorp High School, while Catholic school students went to Countess Moore High School. (When Countess Moore became coeducational in 1969, the school's name was changed to Moore Catholic.) Junior high kids were not exempt as the Berta Dreyfus school in Stapleton was accepting seventh and eighth graders in need of summer classes. A remedial reading program was also available at the school for ninth graders. Youngsters looking to expand their education voluntarily could take a class in genetics at Countess Moore, while others could learn typing or study electronics at Dreyfus.

Something new was taking place when the school year began in 1968. Graduating eighth graders from the previous school year at Prall Junior High would not be staying at the school for ninth grade. Instead, they were on their way to Susan Wagner High School. The Department of Education was beginning to move all ninth grades to high schools. This system meant that some sixth grade elementary students would be shifting to Prall. All sixth grade classes would eventually be moved to the island's junior high schools.

Enrollment at Staten Island's public schools had reached more than 43,000 students as of January 1968. Citywide, school registration stood at 1,109,644. All the boroughs had gained students except Brooklyn. This made sense since so many Brooklyn students were now living on Staten Island. On Staten Island, there were 24,529 students in elementary schools. Junior high schools had 8,086 students, while high schools had 8,961 students. McKee High School had an additional 960 students. This was on top of the student population from the previous year. Another 1,004 pupils participated in special schools for the physically and developmentally disabled, prekindergarten, and nongraded classes.[209]

All were surprised when the Staten Island Federation of PTAs approached city officials during a public hearing on the 1967–68 capital budget in early February 1967. In addition to what was already earmarked for Staten Island, they made another request: an elementary school in the Willowbrook area of the borough. Then they went home—quietly. It was an odd moment for city officials, who expected the usual blazing budget battles provided by the mothers of Staten Island. In fact, council members and members of the Board of Estimate were left speechless. Eventually, PS 54, which would seat 1,200 students, would be built at Sunset and Fields Avenue. It was named after naturalist Charles W. Leng. The presence of PS 54 would alleviate overcrowding at PS 23 in Richmond Town, PS 30 in Westerleigh, and PS 29 in Castleton Corners. According to a PTA study, there were 4,164 new houses planned for Staten Island's North Shore, while the South Shore

Enrollment in Staten Island schools steadily increased after the opening of the Verrazzano-Narrows Bridge in 1964. *New York Public Library.*

would see another 1,100 new houses. These structures were to be completed by 1971.[210]

The portable shuffle was still going strong in 1969 when eighteen additional portables were ordered by the Board of Education. They were mainly for schools in Huguenot and Dongan Hills. Over the years, portables had prevented part-time schedules at a number of schools. But as enrollment continued to climb after the opening of the Verrazzano-Narrows Bridge, classroom space in main buildings was just not available, nor was money for additions to existing schools. Even so, the construction of new homes continued on Staten Island. The PTA begged the board for additional classroom space at PS 5 in Huguenot since several hundred homes were to be constructed in the neighborhood. But there would be no relief. Instead, the eighteen expected portables would provide space for 550 students. If the eighteen portables arrived, Staten Island would be home to seventy-two of these units. To keep schools off part-time schedules, the Board of Education was also redistricting. This process shifted pupils from crowded schools to those that were not nearly as crowded or as overcrowded. The board also rented school space in housing projects, churches, and abandoned schools. According to school officials, portables were the "first line of defense against overcrowding" until new schools were built.[211]

As noted, the PTAs of Staten Island were fierce proponents for building new schools. Both the Federation of PTAs and the local school board stated in 1969 that the borough would need an additional eleven thousand school seats. For this reason, they stressed that six new schools needed to be constructed. Staten Island was set to receive "three of the 45 new schools and additions in the 1970–1971 school capital building program."[212] All three groups—the Federation of Staten Island PTAs, the local Staten Island School Board, and the New York City Board of Education—agreed that a land bank was needed to preserve future school sites as developers gobbled up what was left of available land. The federation stated, "This sensible, logical approach to prevent exorbitant waste of moneys and time for condemnation of privately owned land must be impressed on the budget director and mayor's office so that land-banking becomes an immediate reality."[213] One school board member believed developers with massive plots of property should be required to keep land for future schools, just as they were required to provide streets and sewers for their developments. It was further noted that the shortage of teaching Sisters and Brothers for Catholic schools was increasing the public school population. Parents did not want to pay rising parochial school tuition for lay teachers. It was believed that one thousand children would be transferred to public schools in 1970 owing to this fact.

In 1969, a Staten Island public school was named for an African American. This was a first for the borough. Intermediate School 61 in Brighton Heights was named in honor of William A. Morris, a civic and business leader. Evelyn King, a prominent teacher and coordinator of the new Black studies program in island schools and the daughter of Mr. Morris, hailed the choice. The public rallied around the name, believing that Black students would thereby better identify with the public school system. One local African American youngster from the Dreyfus Intermediate School in Stapleton stated that "Morris fought for 'what is right' and never got discouraged when racists tried to push him out of business. He kept right on going."[214] William Morris established a successful moving company in West New Brighton. As a teenager in North Carolina with little means, Morris decided to move north. His business would go on to be one of the most successful in the moving trade. Morris was also the founder and first president of the Staten Island Van Owners Association. It was Morris who founded the Staten Island Branch of the National Association for the Advancement of Colored People (NAACP).

Monsignor Joseph Riordan, assistant superintendent of archdiocesan schools, stated that "[one] thousand new residents every month, twelve thousand a year or 66 new residents every day for the average school year" were descending on Staten Island. His comment was published in the *Advance* on April 24, 1965, only a few months after the opening of the bridge. Those numbers would only increase. The Catholic school system of Staten Island was experiencing a tremendous upsurge in student enrollment—so much so that a number of Catholic elementary schools and additions were built.

The Assumption School in New Brighton opened its doors before the bridge opened its toll booths. The $650,000 school on Kingsley Place welcomed its first 100 students on March 19, 1962. Workers were still hammering and sawing as the youngsters entered the structure. Their education would come from the Presentation Sisters. When complete, the school would accommodate 450 students from kindergarten to eighth grade.[215]

Monsignor Patrick Gallagher of Saint Patrick's Church in Richmond Town announced plans for construction of an eight-classroom addition to the existing Saint Patrick's School in October 1965. In addition, a convent for the Sisters of Saint Dorothy would be built opposite the school on Richmond Road. The two-story structure would accommodate twelve nuns. The monsignor stated that a building fund was being established to raise $200,000 of the needed $600,000, which would defray some of the construction costs. The parish and its school had grown steadily. In fact, the parish had gone from 450 families in 1962 to 950 families in three years. In the fall of 1965, 587 pupils attended the school.

A new school for Saint Joseph's Church (Rossville) and Saint Thomas the Apostle in Pleasant Plains was to be built. It would stand behind Saint Thomas the Apostle. The eight-classroom building would have 400 pupils. It would be open by September 1966. The parish covered twelve and a half acres and served 550 families. The entire project was expected to cost $600,000. The two parishes were in the process of a much-needed $100,000 building campaign in 1964. Opened in 1848, the Rossville church is the oldest standing church in the borough.

For the school year beginning in September 1965, the number of students enrolled in all forty-five Catholic schools, both elementary and high school, had increased by 1,000 students. Hence, the overall student body for the borough was 22,218. The largest increase was at the high school level since Countess Moore, Monsignor Farrell, and Saint Joseph by-the-Sea were now in operation. Countess Moore officially opened on September 14, 1964, although students were receiving instruction at Saint Sylvester's grammar school in Concord before the Bulls Head campus was ready.

Monsignor Farrell High School in Oakwood became available in 1961, while Saint Joseph by-the-Sea in Huguenot began welcoming students in 1963. The Catholic high schools on the island had 4,466 students, while the elementary schools had 17,752 by 1965. These numbers reflected that the Catholic school system on Staten Island taught 40 percent of the school population.[216] Nationwide, Catholic school enrollment was down. In 1966, there were 58,000 fewer pupils than the previous year. The Archdiocese of New York was an exception. Archdiocese enrollment included students from the Bronx, Manhattan, and Staten Island as well as the upstate New York counties of Dutchess, Orange, Putnum, Rockland, Sullivan, Ulster, and Westchester. The increase in Catholic school attendance in these counties indicated that American families were moving to the suburbs.

All were excited when it was announced that Francis Cardinal Spellman would officiate at the dedication of Countess Moore High School on September 8, 1964. This high school would welcome 375 students that year. The cardinal would also dedicate an adjacent convent that was built for the Presentation Sisters who taught at the school. The campus consisted of eight acres bordered by Arlene Street and Merrill Avenue. Incoming students were part of the freshman and sophomore classes. Full enrollment for all four grades would consist of 500 students.

The *Advance* presented a photo display of Countess Moore High School. The spread showcased state-of-the-art equipment in a language laboratory, while the latest scientific apparatuses were on view in the chemistry lab. Students also had access to a rose garden in the center of the school and a second-floor student lounge. In 1964, Countess Moore had twelve nuns and four lay teachers. The school and the convent cost $2.8 million.

Almost every community in Staten Island had a Catholic Church by the 1960s—sometimes more than one. Catholic families in these communities were willing to financially support churches, schools, construction projects, expansion efforts, and more. But a major problem was facing the archdiocese: a shortage of teaching nuns. Their numbers were expected to plummet. Therefore, the average class size in the Archdiocese of New York was fifty. An increase in the number of lay teachers and improved salaries for these educators caused Catholic school tuition to jump by the start of school in 1967. Both grammar and secondary schools were affected, although not all schools raised their tuition. Rate hikes were as high as five dollars a month, while other schools increased their tuition by one or two dollars per month.

Some parish schools added additional bingo games to the calendar in order to raise money. One Port Richmond parish asked that parishioners raise their weekly church donation by 30 percent in order to support its schools. Increased salaries for lay teachers were part of an effort by the archdiocese to improve educational programs.

After it became legal in 1958, bingo began subsidizing Catholic schools and other not-for-profit organizations. It created a lifeline. Between April 1 and September 30, 1966, the New York State Bingo Commission documented that 27 Staten Island organizations staged 671 bingo sessions. Approximately 19,746 games were played by 193,239 island residents. These groups netted $287,771, while the commission received $9,020 in bingo license fees. Participants split $636,213 in prize money.[217]

Notre Dame Academy on Grymes Hill began planning for new construction during the 1950s. By mid-1963, its expansion fund totaled over $121,000. School officials hoped to raise $150,000. The money would help defray the construction costs for a new wing, a grammar school project, and a previous wing that had been added to the high school in 1959.[218]

The fiftieth anniversary of both Saint Peters High Schools, for girls and boys, was celebrated in 1966. Saint Peter's High School for girls was located on Richmond Terrace in New Brighton, while the boy's school was on Henderson Avenue in Livingston. Originally, boys and girls shared the Richmond Terrace school when it opened in 1916. It was the first Catholic school on Staten Island. In fact, Monsignor Joseph Farrell brought the school to fruition while he was pastor of Saint Peter's Church. Eight girls graduated in the class of 1920. Prior to World War I, it was believed that boys needed only an elementary school education. Beginning in 1943, boys from the grammar school were sent to Henderson Avenue for high school. That building had previously been used by Manhattan College.

The opening of three new Catholic high schools did not provide room for all the arriving children who were registered for a Catholic grammar school education. By 1966, the problem was getting worse since so many parishes were building grammar schools and additions. For this reason, high schools began taking the students who scored highest on entrance exams. This left many students without admittance to a Catholic high school. In addition, no new high schools were to be built. Meanwhile, a grammar school for Saint Christopher's in Grant City, an addition at Saint Teresa's School in Castleton Corners, and other projects would add another 2,400 seats to Catholic grammar schools. High school seats were at a premium.

No one was pleased when rumors swirled that Augustinian Academy on Grymes Hill was closing. Referred to as Augie, the institution was founded in 1899. It began as a college preparatory school and minor seminary for the Augustinian order of priests. The school was experiencing a financial crisis. Many complained that financial assistance had not been received from the archdiocese. In 1966, most of the students were "day hops": they attended during the day and did not board at the school. The declining number of potential priests was affecting the institution as they lived on campus during their studies. After graduation, these students went to seminary schools. It was further rumored that the school and its property were being sold. Some claimed it would become a Catholic retreat. Twenty-nine men participated in the last commencement on June 20, 1969. The land where the Augustinian Academy stood is now owned by Wagner College.

In December 1967 came the announcement that Saint Louis Academy would be closing. Opened in 1904 with twelve immigrant nuns in charge, it now sat on ten and a half acres of land in Pleasant Plains. The school began as an elementary school on Main Street in Tottenville and moved in 1930. The elementary school was phased out in 1966 to enlarge the high school. Through 1962, Saint Louis had also been a boarding school. The Sisters of the Marianites of the Holy Cross staffed the school, but by 1967, there were only seven nuns left. This increased the cost for lay teachers; plus, the school building needed modernization. For these reasons and others, the school shut down. The final commencement at Saint Louis Academy was held on June 13, 1970, when seventy young women received their diplomas. Today, it continues its teaching purpose as the Yeshiva of Staten Island.

Reverend Carl Sutter announced in 1964 that a new wing would be attached to the main building of Trinity Lutheran School in Stapleton. A basement gymnasium, nine classrooms, a repair shop, a medical room, a teacher's lounge, a boiler room, and an administration space would be included. Pastor Sutter expected construction to be completed by the end of the year. Two other Lutheran schools operated at this time, including the Eltingville Lutheran School and Saint John's in Castleton Corners. Both schools had expansion plans in 1966. At this time, Saint John's was considering the location for a new high school. The Eltingville school was also erecting a building that would double enrollment.

By 1966, Richmondtown Preparatory School, a private institution, was open. College prep courses were offered to grades seven through twelve. Located on Richmond Road at Andrew's Avenue, the school had seven classrooms and cost $60,000 to build. It would accommodate seventy-five students, whose parents paid a tuition of $1,050 per year. Alvaro Sanchez of Westerleigh was the owner and headmaster of the school.

Staten Island Academy opened in 1884. The school was operating on eleven Todt Hill Road acres by 1967. A master plan for the school called for the construction of a library, additional laboratories, and a visual arts building with a music room, an auditorium, and a stage. It is the oldest operating school on Staten Island. The academy served as both an elementary and secondary school for students. Parents of the 323 students at the academy paid from $700 for prekindergarteners to $1,200 for seniors. Forty teachers provided education for the pupils, allowing for a low student-to-teacher ratio. On average, 90 percent of students went on to college around 1967. Students were admitted by passing an entrance examination.

Education takes many forms, and the Staten Island branches of the New York Public Library have added considerably to the knowledge of borough residents for many decades. An historic event took place on May 7, 1964, when a modern bookmobile made appearances at the Forest Avenue Shopper's Plaza and at the Midland Beach Playground. The moving library made stops at communities in the borough that lacked a permanent library. It was an important resource for Bulls Head residents who loved to read. The contemporary bookmobile replaced a 1949 model. It was popular and busy. It had not only a fine selection of reading material but also air-conditioning and heat. The first bookmobile on Staten Island debuted in 1922; in fact, Staten Island had the first bookmobile in New York City.

Wagner College, opened in 1883, was originally located in Rochester, New York. It began as a pre-seminary school for Lutheran ministers. In 1918, the school relocated to Staten Island. In an attempt to keep pace with its growing student body, the college embarked on an aggressive building campaign around 1960. In response, a group of modern buildings was constructed. The first new structure was the Horrmann Library, which became available in 1961. By 1966, a new gymnasium and dormitories had been constructed. The college also had a low student-to-instructor ratio of fourteen to one. It sat on a bucolic eighty-six-acre campus atop Grymes Hill, often compared to the grand college campuses of the Northeast. Both the Megerle Science Building and the Spiro Communication Center were built in 1968, while a new Student Union opened in 1970. Previously, a five-floor, $3.5 million dormitory with 604 beds had been constructed in 1964. With all these improvements, Wagner College saw enrollment rising to three thousand students by 1970.

The Wagner College campus atop Grymes Hill (seen here circa 1930–1950s) is often compared to the grand college campuses of the Northeast. *New York Public Library.*

In 1956, Staten Island Community College opened. Fourteen faculty members taught 113 students liberal arts, preengineering, and medical, mechanical, and electrical technology. The first graduating class of 1958 accepted their diplomas at a ceremony held at Sailors' Snug Harbor on June 9 that year.

By the mid-1960s, tens of thousands of students were enrolled in two-year degree programs at the community colleges of the City University of New York (CUNY). For those Staten Island students who wanted to continue their education after graduation, a population that was skyrocketing in the 1960s, Richmond College was founded. This was fortunate since junior and senior slots at most four-year colleges were filled by their own students. Richmond College was an "upper divisional" or senior college located in Saint George. It was the third college of its type in the country and the first in an urban area when it was founded in 1967. Interdisciplinary work was stressed. A radical feature of the school was its pass-fail grading system, intended to support students in learning for learning's sake rather than pressuring them to make the grade. It was said at the time that Yale copied the system and it became a trend at other colleges. Interestingly, students were not happy with the system. They wanted grades, especially if they planned on going to graduate school, where passing or failing meant nothing to administrators. Another aspect of Richmond College was the school's willingness to explore programs of study considered radical at the time, including computer science and urban studies. As one college scholar said in 1965, "Someday we'll be able to transfer information within seconds between colleges by using TV or transmitting facsimiles....They call it 'information retrieval.'"[219]

Construction of Staten Island Community College at Sunnyside was well underway by late 1965, when the City Planning Commission allocated $1,421,150 to complete the project. The entire task was expected to cost $13 million.[220] As Staten Island's population grew, the college would be utilized by thousands of individuals, including "traditional"-aged college students and older returning students seeking a college degree to improve their job prospects or advance at their current jobs.

Staten Island Community College would house a nine-hundred-seat auditorium. In addition, the fall of 1964 saw plans for a television studio and a radio workshop. The radio workshop would evolve into the college's own radio station, WSIA. Planners believed that the student body of the community college would consist of 2,500 students by 1966.[221]

In 1976, Staten Island Community College and Richmond College merged, forming the College of Staten Island. There were two campuses at this time: one at Sunnyside and the other at Saint George. The student population began meeting at a modern, coalesced campus that opened on the site of the former Willowbrook State School in 1993.

Notre Dame College, a Catholic undergraduate college for women, received its charter in 1933. It had thirteen students that year. It was the first school to offer a college degree to women on Staten Island. In 1955, the college doubled the size of its property with the purchase of the Herbert Gans estate on Grymes Hill. September 1964 saw the largest freshman class ever enrolled. Mother Saint Egbert, president of Notre Dame College, stressed that the opening of Mahoney Hall in 1962 had benefited the college with its modern laboratories, faculty office and research space. An addition to Lavelle Hall was to be completed during the summer of 1964. Another structure scheduled to open in 1966 would provide a modern student dining room, a contemporary language laboratory, student lounges, lecture rooms, faculty offices, and a chapel wing. Tuition in 1964 was $850 per year.

During the spring of 1966, school president Mother Saint Rita Marie of Notre Dame outlined a ten-year development plan. Noting that all were caught up in the excitement of Staten Island's future, she called for school expansion. Mother Saint Rita believed it was necessary to study enrollment trends, enrollment projection figures, student-faculty ratios, faculty needs, and the college's faculty recruitment processes. She called for improved library conditions, a college chapel, a residence for religious faculty, and an academic building that would include classrooms, student space, a bookstore, and a cafeteria. To meet these needs, first, a consulting company was brought in to examine other nearby colleges. The second phase was to involve the college becoming more involved with the community. Tutoring high school students, providing a reading clinic for the public, and making the college available for seminars was suggested. Recruitment of qualified freshman from local high schools would also take place, as would expanding the physical facilities of the site.

The following year, Notre Dame College made an unexpected move when they hired Dr. Carson W. Veach to serve as academic dean. Not only was he a member of the laity, but he was also an Episcopalian. Veach was expected to focus on expanding the college, emphasizing curriculum, and promoting a close relationship with the lay teachers, who were now two-thirds of the teaching staff.

As the 1960s marched on, it was becoming difficult for smaller colleges to survive. Terence Cardinal Cooke, then archbishop of the Archdiocese of New York, asked the Vincentian Fathers of Saint John's University in Queens to integrate Notre Dame College. The cardinal believed it was necessary for Catholic higher education to be available on the island. On January 27, 1971, the New York State Board of Regents approved the consolidation and the Staten Island campus of Saint John's University. Notre Dame College on Staten Island ceased to exist. Classes began at Saint John's University on Staten Island that fall. Undergraduate degrees in liberal arts, business, and education were available to students who enrolled and successfully completed course requirements. Saint John's on Staten Island met the needs of not only the numerous Catholic college students living in the borough but also the new residents who had moved from Brooklyn and Brooklyn residents who traveled across the bridge to attend classes.

The Jewish community was growing exponentially as the move from Brooklyn to Staten Island continued. The oldest borough congregation, B'Nai Jeshurun, was eighty-seven years old in 1971. Their Victory Boulevard building had been in use since 1891. For expansion purposes, it had been renovated in 1920. After the bridge opened, it was far too small for the number of people moving to the borough. By 1967, the congregation had 180 families, compared to 120 families in 1920, and the numbers were rising. The congregation had resorted to using the gymnasium of the Jewish Community Center to accommodate all who attended services on High

Holy Days. United Synagogues of America determined that Staten Island's Jewish population would double by 1969. Congregants of B'nai Jeshurun were involved in many aspects of synagogue life, including the Sisterhood and Men's Club, the United Synagogue Youth, the Adult Institute of Jewish Students, and the Hebrew day and Sunday school. The *B'nai Jeshurun News* was a popular publication with members. B'Nai Jeshurun began moving into the former Steers and Steers Funeral Home building on Martling Avenue in West New Brighton in October 1971. The former synagogue at 199 Victory Boulevard in Tompkinsville was put up for sale. The building on Martling Avenue was to be used as a temporary house of worship until a new temple was built on the property.

Ground was broken on August 17, 1969, for a new synagogue in Willowbrook. Young Israel of Staten Island was an Orthodox Jewish synagogue on Forest Hill Road. The ceremony featured three politicians running for mayor of New York City: John Marchi, Mario Procaccino, and Abe Beame. The congregation began in 1965 when five families met and planned for the future. Services began in 1967. Sixty families comprised the congregation when ground was broken. They met in members' homes.

Chapter 14

URBAN RENEWAL ON THE SOUTH SHORE?

The Annadale-Huguenot area on the South Shore of Staten Island held approximately 1,083 acres. A large percentage of this land was vacant in 1964. The entire area surrounding Annadale-Huguenot was referred to as South Richmond. Within this location, 8,300 acres was still considered to be vacant in the early 1960s. According to the *New York Times*, South Richmond was the acreage south of Great Kills Park and the Fresh Kills Landfill, excluding the built-up areas of Eltingville, Tottenville, and Great Kills.[222]

The Annadale-Huguenot tract was marked for urban renewal by the New York City Planning Commission in 1962. The commission was attempting to "plan the growth of an area before houses crowd the land, and streets that are now just marks on maps become concrete or tar surfaces."[223] The definition of *urban renewal* is "the rehabilitation of city areas by renovating or replacing dilapidated buildings with new housing, public buildings, parks, roadways, industrial areas, etc. often in accordance with comprehensive plans."[224] Even before the plan was realized, the Staten Island Real Estate Board and the Staten Island Home Builders Association recorded their opposition. It was not long before the residents of Annadale and Huguenot and the majority of Staten Islanders followed suit. In fact, many members of the public were outraged. To them, urban renewal meant apartment buildings, Big Brother government, and African Americans. One builder later said, "They left South Brooklyn when Negroes moved in. They escaped here. They don't want any more people on the island."[225] It was believed by many that African Americans being displaced by highway construction in the Bronx would be moved to the island's South Shore by the city.

The Staten Island Real Estate Board was "against having the slum clearance goal of urban renewal diverted to development of open land areas." They believed this was the beginning of "prevent[ing] development of the Island and other suburban areas with private homes." The board made clear that a "plan of action [was] being formulated" and would be presented at a public forum that was to be held in late February 1963 before the New York City Planning Commission.[226]

The Staten Island Home Builders Association stated that the Annadale project was a "grievous mistake." The association believed growth would be slowed by six to ten years if urban renewal was begun. Instead, it called for private development within the next two to four years. The association noted that this would lead to an increase of New York City's tax rolls.[227]

Many island groups announced their displeasure regarding the Annadale-Huguenot urban renewal. Along with the previously mentioned fears, many individuals believed their properties would be condemned. This appears to have been a result of the displacement that occurred from the installation of the Staten Island Expressway.

An editorial in the *Staten Island Advance* showed support for urban renewal in Annadale, Huguenot, and the surrounding area. While it was noted that the term *urban renewal* was not appropriate for an area of open space, the editor opined that a well-thought-out plan could lead to development that was pleasing for both current and upcoming residents of South Richmond. It was stressed that "advance planning is far cheaper than rehabilitation." The editorial noted that the city owned considerable acreage and "thus, the city still holds the reins on development," something the public was beginning to clamor for. The proposal would cover sensible plans for housing, sewers, drainage, schools, recreation, traffic, and transportation. As stated by the *Advance*, "It takes in every major factor that makes a community tick happily."[228]

On February 25, 1963, a hearing sponsored by the Planning Commission was held at City Hall to discuss urban renewal of Staten Island's South Shore. Approximately one hundred residents made an appearance. Most were against the plan. Some of the opposition threatened that if the commission approved the plan, it would be brought to court. Interestingly, it was noted that a plan had not yet been prepared. The commission explained that the point of the hearing was to determine if the 1,083-acre tract was suitable for urban renewal. If it was, an application would be made to the federal government for funds to launch a study and develop a strategy for implementation. The hearing did little to allay public fears of land condemnation even though officials stated that every effort would be made to allow landowners to retain their properties.

Borough President Albert Maniscalco supported the urban renewal plan for the Annadale-Huguenot area—initially. On September 9, 1963, a hearing on the plan was held at Borough Hall. Hosted by Maniscalco, the hearing was rowdy at times; some attacked the program as "socialistic and an invasion of private rights." Maniscalco was in the process of reconsidering his early support and wanted to examine the plan which he deemed "vital" to the interests of the borough. Most of those concerned with the urban renewal plan were homeowners within the tract, while those who spoke in favor of the plan lived elsewhere. The Staten Island Real Estate Board, the Chamber of Commerce, and members of the building industry sent representatives to voice their opposition. The Liberal Party of Staten Island and the Staten Island Citizens Planning Committee spoke in favor of urban renewal for Annadale-Huguenot. More than 250 individuals were in attendance. The urban renewal plan was approved by the Planning Commission with the result that the City Housing and Redevelopment Board was to draw up a design for a "model community" bounded by Arden Avenue, Amboy Road, Huguenot Avenue, and the Raritan Bay.[229] Many of the opponents called for private builders to develop the land. Proponents stressed that this opportunity for sound planning should not be lost.[230] The borough president listened to the majority of his constituents and called for the derailment of the urban renewal plan for the Annadale-Huguenot tract.

While the plan's demise was slowly winding its way to completion, the price of real estate in Annadale and Huguenot was escalating. Attorney and land broker Guy Molinari, who would be borough president of Staten Island from 1990 to 2001, announced in early November 1965 that property in the urban renewal location was "unofficially assessed recently to be worth about $50,000 an acre. The cost of a building lot would be approximately $7,000 and potential density would be seven houses on an acre."[231]

Building on Staten Island had actually slowed by 1966 owing to high interest rates and the demise of the Annadale-Huguenot plan. But in 1967, the rates went down, and a court ruled that developers could build on the Annadale-Huguenot properties. That same year, Mayor Lindsay earmarked $200,000 to draft a development plan for the Annadale-Huguenot area. City planners soon stated that the best alternative was a quasi-public development corporation that would buy up land in all South Richmond. The corporation would then change plot sizes and street patterns, and the land would be resold to responsible banks or developers for future use. Builders argued that the differing houses on larger plots would be more expensive for buyers and the higher prices would block many from purchasing a home.

In 1967, Raymond & May Associates released alternative plans for the Annadale-Huguenot area. They envisioned the city selling land to private developers who, along with the city, would share the cost of sewers and roads. All the plans presented would save most of the 500 homes in the Annadale-Huguenot area, preserve streams and trees, increase beach usage and parkland, and move the shorefront parkway away from the beach. The community would be largely middle income. There would be no provision for low-income housing. Single-family homes would number 1,000, while attached townhouses for 3,750 families, an eleven-story apartment tower, and garden apartments that would hold 400 families would be constructed. Housing would be grouped in small clusters around an open community-owned space. Greenways would connect the clusters.[232]

Residents of the area were not pleased. So again, in January 1969, Mayor Lindsay offered another proposal that would create a model community with a rural flavor. Housing, commercial, and recreational facilities for 20,000 people on the 1,080 acres of largely undeveloped tracts in the Annadale-Huguenot area would be created. The land would be heavily wooded. This proposal called for 5,350 dwellings, including 1,400 single detached houses and 2,660 units in townhouses. The remaining residences would be included in apartment buildings. Streams, trees, and natural topography would be maintained and a lagoon and marina established, and 85 percent of existing homes would be preserved. But builders and many residents were opposed to any plan that restricted development. Of course, many of the residents were frightened at the prospect of having their house condemned. Borough President Robert Connor opposed the proposal and called it urban renewal. Connor believed that the area was "neither urban, nor [did] it need renewal."[233]

In 1970, the Rouse Corporation released a plan for South Richmond—the land south of Great Kills Park and the Fresh Kills Landfill, excluding the built-up areas of Eltingville, Tottenville, and Great Kills. This proposal called for spending $6.5 billion to build a "new city within the city."[234] South Richmond consisted of 10,600 acres, 7,300 of which were vacant or undeveloped; 45 percent of the expanse was city owned. The report recommended, as others had in the past, that fill be used to create an additional 3,300 acres to increase the shoreline along the Raritan Bay. This move would create more city-owned vacant land. The plan called for a population of 450,000 who would work and live within the borders of South Richmond. A public corporation would manage the area. It would be formed by special governmental legislation. The corporation would have many responsibilities, including administration of a land bank and establishing a plan for development.[235] In addition, the corporation would have the power to "erase obsolete boundary and parcel lines and clear up" title questions; create usable parcels for public and private development; "raise large amounts of long-term funds at the lowest possible interest rates"; "install or have installed a wide variety of land improvements including roads, utilities, shoreline improvements, and public and semi-public buildings"—including schools, parks, and pools; and "function continually in the face of shifting political winds."[236]

This development plan was called the South Richmond Project. "Pre-servicing" would be a feature of the plan. This would entail having the necessary elements for living in the houses and the community before new residents arrived.[237] Many of these elements were lacking in communities that were built before and after the opening of the Verrazzano-Narrows Bridge, including but not limited to streets, sewers, schools, and recreational facilities. Underground lines for telephone and electricity were suggested.[238] To help residents understand the Rouse Plan, which had already been presented by the company to almost one hundred community groups and organizations by this time, the *Advance* presented a four-part series describing the proposal in February 1971.

Reston, Virginia, a specially designed community near Washington, D.C., was often referred to as a model of how the open areas of 1960s Staten Island should have been laid out. In fact, the South Richmond Plan took many ideas from the Reston concept. Simply put, Reston was a well-planned community that brought open space, residences, and commercial interests together. It was founded by Robert E. Simon (the Reston name begins with Simon's initials), whose goal was to provide "maximum opportunities for 'physical and mental exercise' and a safe and beautiful community to meet the needs of people who are facing diminishing work weeks and increasing leisure times."[239] In fact, these words were spoken to a group of 350 people at a Tavern on the Green luncheon on November 15, 1966. The event was the Fifth Annual Buildings Awards, sponsored by the Staten Island Chamber of Commerce. Reston mixed detached houses, townhouses, and high-rise apartments for a population of approximately 75,000. Simon advised that builders, planners, politicians, and the community consider recreation (in other words, plan for parks, walkways, and bicycle and bridle paths); eliminate the City Planning Commission in favor of a small system of offices (Simon was incredulous that a group in Manhattan could deal with all the differences in the five boroughs); obtain federal funds for planning; have developers create vest pocket parks; study the playgrounds of England and Scandinavia; and condemn land before proposed uses and then sell it back for specific uses.[240]

The Rouse Plan or plan for South Richmond cost the City of New York $715,000.[241] It was soundly rejected by island residents and by the New York State Assembly in 1973. Instead, the City Planning Commission adopted traditional zoning plans.

Back in 1967, the city proposed a zoning change that permitted developers to place houses close together and create a private park on the remaining green space. This would also preserve natural features. The proposal provided financial incentives to create these "cluster houses." When Mayor Lindsay spoke at a luncheon hosted by the Builders Association at this time, he implored the men to consider this suggestion. They were not well received.

History was made when a planned development at Arden Heights was announced in 1970. It was, up to this time, the largest single-family home development within the City of New York. Mayor Lindsay called the $100 million private venture the "first large scale example of 'planned unit development' of the kind that his administration has been advocating for residential areas in the outer boroughs."[242] The site, called Village Greens, would "consist of nine 'Villages' each containing 225 attached townhouses in rows of seven to twelve arranged around green spaces." It was bounded by Arthur Kill Road and Arden, Huguenot, Rosedale, and Vespa Avenues. No through roadways would be installed in this 160-acre area. The site developer paid $9.5 million for the property. Loew's Corporation, the hotel and theater chain, along with J.H. Snyder Company, a construction firm from Los Angeles, were jointly sponsoring the project. Construction began on March 4, 1970. Occupancy would begin by the end of the year. A three-bedroom unit would sell for $30,000, while a four-bedroom dwelling would be available for $38,000. In this type of cluster community, homeowners gave up the traditional forty-by-one-hundred-foot plot for a modest-sized front and backyard. In exchange, they would have access to a commonly owned green space and recreation area that had two pools, tennis courts, a clubhouse, and a playground. The builders hoped to instill a natural New England feel, so they used white clapboard and stone on the facades. A public school was eventually built on the property, as was a six-and-a-half-acre shopping center.[243]

Thousands of trees were destroyed before and after the opening of the Verrazzano-Narrows Bridge. Countless articles documenting their demise were published for years. Even the borough president could not stop the felling of trees. In 1965, Albert Maniscalco, Councilman Edward Curry, and the residents of New Dorp tried to stop the carnage. Twenty-three shade trees on the south side of Bryant Avenue were cut to the stump and hauled out for a sewer project.[244] The sewer was installed to reroute the Moravian Brook, which then flowed through an open-walled course. Another smaller private sewer was also being built. Both lines would run from Richmond Road to the Staten Island Rapid Railroad tracks. A suggestion that the sewer line run through Locust Avenue to save the trees was presented, but this was deemed to be too expensive. Other proposals were also disregarded. In response, the company removing the trees said it would plant nineteen or twenty trees at the company's expense.

Borough President Robert Connor called for an official style of streets in January 1968 but only around the area of the Richmond County Country Club and Dongan Hills. Connor stated that the streets would be "in line with model cities, [a] planned community concept." The borough president wanted "narrow, sidewalk free" roads that would be developed only at their existing widths, not the generally required widths. Connor believed this would "retain slopes, lawns, and natural assets—to preserve peaceful residential communities."[245]

Chapter 15

AFTEREFFECTS

FROM TRANSPORTATION TO TRAFFIC

On Easter Sunday 1969, a record one hundred thousand cars crossed the Verrazzano-Narrows Bridge. Traffic was backed up on the Belt Parkway all the way to the Coney Island exit. Not only were people visiting family in Brooklyn and then coming back to Staten Island, but many were also looking at the new homes that stood side by side in the housing developments along Hylan Boulevard. One former Brooklynite was quoted as saying, "No, it's not heaven. But when robins start nesting in my apple tree, I forget that my cesspool doesn't work and that I can hear my neighbor's mother-in-law yelling at him." When one policeman was asked by an off-island visitor where the nearest farm was, he replied, "Lady, you're about five years too late."[246] Actually, there were at least four. The Mohlenhoff, Bifulco, Chrampanis, and Gericke farms were still in business, as were a number of smaller operations.

Many changes had occurred on the island since the bridge was declared open on November 21, 1964. Some of the upheaval came in the area of marine transportation. The Brooklyn and Richmond Ferry Company commenced service between Sixty-Ninth Street in Brooklyn and Staten Island on July 4, 1912. In 1939, Electric Ferries, aptly named because it built and used diesel-electric boats, began operating this line. Electric Ferries did not renew its franchise in 1954. These boats landed at a small dock in Saint George. In 1961, 2,268,613 passengers and 2,658,334 vehicles made the crossing. These ferries were now operated by the 69th Street Brooklyn Ferry Corporation. The company's profits came from automobiles, not foot traffic. Before the Verrazzano-Narrows Bridge was complete, these ferries were in demand, especially during the evening on summer Sundays. It was not uncommon to see long lines of cars waiting for boats on the Staten Island side in the mornings. This view would be repeated in the evening in Brooklyn. Sitting in sweltering cars (most autos lacked air-conditioning), passengers waited while ferries filled up and departed and then filled up again and departed. It was distressing. These ferries were small, so there was a lengthy delay in getting home. With the opening of the bridge, this ferry line changed radically: ridership fell so quickly that it closed four days after the bridge opened. The final ferry ride on November 25, 1964, was odd, to say the least.

The night was heavy with rain, wild and windswept. The last little boat between Staten Island and Brooklyn had members of the Steamship Historical Society of America onboard plus a number of offbeat young folk. The latter group was distributing copies of *The Ferry, the Bridge and the People: A Eulogy to the 69th Street Ferry*. As they distributed their publication, they sang, "We love you ferry, honest we do." Police officers closely monitored the situation. They wanted to prevent life preservers, benches, signage, and anything else that might not be theft-proof from disappearing. After the last boat left Brooklyn, a desperate few scrambled to grab the coveted

The Saint George landing of the 69th Street Brooklyn Ferry Corporation, seen here circa 1950, was adjacent to the Staten Island Ferry Terminal. *New York Public Library.*

lifejackets. Many of these souvenirs were quickly thrown into car trunks. As the boat made its way to Staten Island, the young ones once again sang, "We love you ferry, honest we do." All of this led one soon-to-be-unemployed deckhand to muse out loud, "Why didn't all these guys show up earlier and keep us in business?"

Even though many commuters would now be driving over the Verrazzano-Narrows Bridge, officials realized the borough's population increase meant that additional Staten Island ferryboats would be needed. Fortunately, three new ferryboats were being assembled. They would be members of the Kennedy class. The *John F. Kennedy*, the *Governor Herbert H. Lehman*, and the *American Legion* would serve the city for many years.

In 1961, ten boats moved 20 million passengers and 1.5 million cars. On a typical working weekday, 40,000 individuals were ferried back and forth to Manhattan. The fare was still a bargain as it cost only the "world-famous nickel." A ride on a city bus or the subway at this time cost fifteen cents.[247]

Not only did the opening of the bridge allow more cars to pass into Brooklyn, but it also enabled express bus service between Staten Island and Brooklyn to begin on November 3, 1965.

The Staten Island Rapid Transit passenger line from Saint George to Tottenville was not faring well in 1964. Ownership issues on the aging system were a concern, and modern railcars were urgently needed. It would be years before they would arrive. Fortunately, the Transit Authority was scheduled to receive 215 buses for its nineteen Staten Island bus lines.

The roads of Staten Island have always been a cause for concern—some of them for centuries. Sections of Amboy, Richmond, Arthur Kill, Manor Roads, Richmond Avenue, and Richmond Terrace date to 1797 and earlier. Some were former Lenape paths. By the 1950s, these roads were abominable owing to narrowness, curves, and lack of upkeep. Even as ninety-five red, white, and blue signs began pointing to the Verrazzano-Narrows Bridge, jarring potholes welcomed the wheels of the Brooklynites' cars. Some of these roads would remain in this condition for decades. A $5 million project to improve Hylan Boulevard as a viable connection between the Verrazzano-Narrows Bridge and the Outerbridge Crossing did begin in 1964.

Traffic was a major concern after the opening of the bridge. The roads of the island were barely accommodating vehicles before the great span opened. As bottlenecks and headaches ensued, changes needed to be made. For this reason, an announcement was made in early 1965 that Victory Boulevard would be expanded to six lanes from Willowbrook Road to Richmond Avenue. The plan called for the boulevard to be widened to sixty feet. The existing road was anywhere from thirty to forty feet wide. Funds were also being requested to widen Victory Boulevard from Clove Road in Sunnyside to Forest Avenue in Silver Lake.

By 1968, an 8 percent increase in Staten Island traffic had been realized. This was double what the other boroughs were experiencing. Traffic numbers were projected to rise for the next three to five years. They would, of course, be increasing for the next sixty years.

Drivers were bogged down on Bay Street in Stapleton in 1968. In fact, traffic on Bay Street had risen by 50 percent since 1964. This held true for sections of Amboy Road in Oakwood. It was hoped that the opening of the West Shore Expressway would alleviate congestion. It was further hoped that the highway would be a boon to industry. The *Advance* noted that 1.3 million feet of line markers had been painted, 3,700 new traffic signs had been installed, and 12 new traffic lights had been introduced in 1967 in order to upgrade driving conditions in the borough. Additional "No parking," "No standing anytime," and "No stopping" signs were expected for the future.

There were many who believed that the Verrazzano-Narrows Bridge would solve the problem of industry leaving Staten Island. Its modern span and expressway would allow trucks to zip over to New Jersey, Brooklyn, Manhattan, and beyond for deliveries. But manufacturers were not interested in improving the remaining factories or establishing new ones. As one businessman stated, "It's a cold, hard fact that the same stringent building and license requirements apply here as in the rest of the city. A company that's thinking about making a move has good reason to feel it would get a better deal if it moved just a little bit further, as into New Jersey."[248]

Chapter 16

RISING CRIME

The rise in Staten Island's population led to a rise in crime. Therefore, law enforcement looked at the existing crime-fighting infrastructure of the borough. By 1965, Staten Island had station houses at Saint George, New Dorp, and Tottenville. There were six hundred police officers. The police commissioner penned an *Advance* article on April 24, 1965. The headline read, "Police Force Growing Along with Borough." He noted that traffic problems were a big concern for the police department. Staten Island presented different circumstances than the other boroughs: a speedy increase in the number of motorists, narrow streets, hilly terrain, and inadequate off-street parking facilities. The commissioner called for additional traffic patrol assignments to meet the increase in cars. He also stated that the emergency service squad based in New Dorp had the latest in emergency vehicles. Motorcycle and mounted patrolmen were also assigned to parks and pathways.

A week later, the commissioner promised additional officers and equipment to combat Staten Island's increased crime. Some of this crime was related to an increase in illegal gambling, which the commissioner vowed that the police department would monitor. He further stated that plans were being made for a new Tottenville station house. Noting that 175 new radio cars had been ordered—no doubt for the entire city—he said there would be an upturn in the number of patrols, especially along bus routes. Additional officers were also being recruited, but it was difficult to attract potential officers. It was believed that young men preferred other professions that were more to their liking. To entice new officers, the pay was rising from $7,000 annually to $9,000 after three years. The department was also providing special training to police and attendance officers to fight the rise in narcotics.[249] Crime in the entire city had also risen between 1963 and 1964.

Islanders were taking measures to combat lawbreaking. One Todt Hill resident pleaded with his neighbors to fight the crime wave. One break-in a month was occurring. In early June 1965, $4,000 worth of valuables was swiped and then removed from the scene with the homeowner's car. In order to invigorate the old Todt Hill civic association, this gentleman became its president. He was able to encourage one hundred families to join the association. The patrol consisted of men at night and women during the day. Burglar alarm installation was thriving. The patrol had already interrupted one robbery. Luminous bumper stickers indicated to intruders that they were being monitored.

Records show that twenty-five burglary complaints were made to the Tottenville Precinct while ninety-five were documented at the New Dorp Precinct in 1965. The Saint George station house did not have figures, but one police official noted that crime had increased since the summer of 1964. He believed that the swelling population had given rise to an increase in businesses and that this rise was in proportion with the rising crime rate. Randall Manor was most targeted. Officials theorized that drug addicts were stealing items for quick cash. Televisions, cash, small appliances, jewelry, and clothing were preferred. It was noted that youngsters between sixteen and twenty years old were becoming more involved.[250] Police believed the rise in crime was only temporary.

Richmond County was the only borough without a jail by 1964. The corrections commissioner informed the Planning Commission that a jail in Staten Island was a "must" because of the opening of the Verrazzano-Narrows Bridge and the resulting population boom. She noted that juvenile delinquency was already rising in the borough. The old county jail at Richmond Town was closed in 1954. Owing to this situation, island prisoners were lodged in other boroughs and transported back and forth to Richmond for court appearances. No specific site for a jail had yet been selected. District Attorney Braisted echoed the call in 1965. He noted the correlation between the increasing population and the rise in crime. Braisted observed that it was inconvenient for lawyers, families of the incarcerated, and correctional officers as they were tasked with transporting prisoners around the five boroughs. In addition, the other jails were filled to capacity.[251]

A rash of attacks on women in Midland Beach led to the establishment of a safety committee whose members patrolled the streets to thwart attackers. The Midland Beach Progressive Association first became concerned in early 1965 when a young girl was approached by a man in a car at the corner of Moreland Street and Greeley Avenue. The man jumped out of the auto and assaulted her with a knife. He took off when another automobile approached. On March 12, 1965, a driver purposely hit a female pedestrian on Olympia Boulevard. A passenger in the car then attempted to grab the woman and put her in the automobile. The community was frightened, especially after dark. They called for another police precinct on the island and additional street lighting. The Midland Beach Progressive Association had approximately fifty members.[252]

In 1964, 900 burglaries took place on Staten Island—a 25 percent increase from 1963, when 781 burglaries were documented. These occurrences and resulting financial losses led insurance underwriters to announce that burglary of both homes and businesses would continue to increase, thus leading to an increase in insurance premiums.

In the early spring of 1966, the Richmond County District Attorney stated that the borough required more policemen. He blamed the growing population and the bridge. He believed the span provided easy access to Staten Island for criminals from the other boroughs.

By 1965, three alarm companies had opened in the borough. They were thriving. Most homeowners were having alarms installed for peace of mind. They did not want their personal items being stolen or destroyed. One company noted that 80 percent of homeowners installed an alarm after their house was vandalized.

Increased criminal activity led the police inspector to advise residents that they should install a chain on their front door so intruders could not force their way in. The inspector stressed that homeowners should always ask for identification when a stranger knocked on the door and requested entrance. He noted that utility workers, city employees, and many business representatives always carried proper identification. He advised residents to call the police if suspicious or unknown individuals were spotted while also jotting down their license plate numbers. He further advised, "Women alone should consider buying a dog." The inspector went on to advise leaving lights on when no one was home and not mentioning extended absences from one's home to strangers. As the borough changed, residents needed to also change. The inspector's tips were included in the publication "Guidelines for Protection of Self and Property."[253] This was quite a turnaround for island natives who generally left their doors unlocked and their car keys in the ignition.

Parks were also under assault. The Westerleigh Improvement Society went before the City Council to present a petition calling for improvements at Westerleigh Park. Residents were horrified by the rampant use and sale of drugs, including heroin, in the park. In addition, the society called for the repair and restoration of restrooms and the park's attendant room. Both were located in a bandstand that was damaged by fire in 1967. They called for a parks employee to be assigned to the site and a nightly police patrol. They complained that drag racing, offensive teenagers, and indecent acts were prevalent at the park. Several days after they submitted the petition, an officer began patrolling Westerleigh Park.

As the population expanded, the police force needed to be bolstered. In 1967, plans were put in motion to modernize and enlarge the department citywide. The result was termed "instant cop." This individual would be a patrolman on a motor scooter. He would be equipped with a walkie-talkie to contact the local precinct. Thirty policemen would be stationed on the island. A department spokesman declared that the borough needed two policemen for every one thousand individuals. "Using this as a guide, the Island, with [an expected] population of a half a million, would require a one thousand man force."[254] It was believed that police would soon be assisted by electronic communications and computers.

As both the population and crime increased, so, too, did local court schedules. The need for additional judges, courtrooms, and court personnel was mounting in 1966. More residents meant not only more traffic accidents and infractions but also an increase in car theft. At this time, the county courthouse on Richmond Terrace was home to the supreme court, the surrogates' court, the county clerk's office, the district attorney's office, and other agencies. Some file cabinets were stacked twelve feet high, while many official records were stored in the attic.

Family court was located at Academy Street and Richmond Terrace. A criminal court was in Stapleton, and a civil court stood in West New Brighton; both needed additional space. Experts called for using the soon-to-be-vacated U.S. Coast Guard base. One judge said a civic center should be built for additional court facilities. He believed it was imperative that it be completed in five years. Otherwise, he warned, Staten Island would face a horrendous situation.

In response to the rising crime rate, the police department added fifteen radio cars to its fleet of thirty-two in 1967. Five of the vehicles would be

As the population of Staten Island increased, so too did the court schedules of local courthouses like the criminal court in Stapleton, seen here circa 1930–1950s. *New York Public Library.*

assigned to sergeants who would be on constant patrol. Most of the vehicles would be assigned to the 120th Precinct in Saint George. Seventy-five additional patrolmen would be added to the existing force, increasing the number of borough police to 675.[255]

In 1967, sources stated that felony crime on the island had increased by 57.6 percent from the previous year. Richmond's residents were beginning to realize that the bridge brought an increase in the kind of problems that had rarely occurred before it was built. Burglaries had risen an astounding 84 percent, robberies were up 55 percent, grand larcenies were up 44 percent, car thefts were up 38 percent, and felonious assaults were up 67 percent in 1966. Some officials claimed the rise was due to a recent change in compiling crime statistics.[256]

In 1967, car thefts had nearly doubled. As such, police began a "Lock Your Car" campaign. The program was sponsored by the Richmond County Insurance Agents Association and the National Auto Theft Association. Police surveyed the Forest Avenue Shopper's Plaza on October 1, 1967. During a two-hour period, they checked 565 cars; 406 were unlocked. Auto theft was a crime of opportunity.[257] At the time, locking a car had to be done with a key. Sometimes, the door key was different from the ignition key, which many saw as an inconvenience; thus, they left automobiles unlocked and vulnerable to opportunists.

Owing to an increase in traffic infractions, it was reported in 1966 that borough insurance rates would be rising. The rate for an unmarried male under twenty-five in 1955 was $153.60. The "skyrocketing" rate for the same individual in 1966 was $287. One past president of the Richmond County Insurance Agents Association directly blamed the bridge and the expressway. He further stated that the increased cost was related to inexperienced young drivers.[258]

Chapter 17

HEALTHCARE CONCERNS FOR AN EXPANDING POPULATION

Near the end of 1958, local physicians cautioned that a serious problem existed on Staten Island: The borough lacked hospital beds. One of the first casualties of the planned Clove Lakes Expressway (Staten Island Expressway) was the Sunnyside Hospital. Opened in 1940 on Little Clove Road, it would be demolished to make way for the highway. Forty practicing borough physicians were so horrified that they petitioned the Triborough Bridge and Tunnel Authority to alter the highway's route. They noted that the hospital's demolition was a problem for those who already used it as well as for future island residents. Owing to the shortage of hospital beds in the borough, they believed Sunnyside Hospital was irreplaceable. The doctors applauded its central island location. Even so, Sunnyside Hospital closed on September 4, 1962. Leading up to its closure, staff had been busy placing patients elsewhere as it was impossible to maintain a healthy environment owing to the noise and dust emanating from the highway's construction. Doctors' Hospital opened on Targee Street in Concord on June 21, 1963. Some viewed it as a replacement for Sunnyside Hospital.

Physicians were not as alarmed about the inadequate number of patient beds on the island once Doctors' Hospital opened. The situation was not resolved, but it was not as urgent. By 1965, the hospital had added 116 beds. One negative result of the alleviation of the bed shortage was that no general hospital would be established on the island. For this reason, the hospitals of Staten Island were urged not to duplicate services. Coordination would be key. Joint planning was needed. The Review and Planning Council of Southern New York also noted that the island had a large population of children under fifteen as well as married women of childbearing age. For these reasons, maternity and pediatric care would remain critical to Staten Island's health care system.

The new year 1964 brought news that Richmond Memorial Hospital in Prince's Bay was expanding to meet the needs of a changing borough. The number of admissions, emergency visits, births, and clinic visits had all increased in 1963. Only the number of patient days and the average percentage of occupancy had declined owing to improved medicine, treatment, and facilities. Even so, the hospital's president announced plans to build a one-story addition to the existing south wing. A new heating plant, a new dietary department, and extensive first-floor changes were listed as essential to increase patient capacity. Administration also called for a nursing home and a new third-floor wing for maternity beds.[259]

The following year, Richmond Memorial Hospital's president discussed a number of accomplishments, including the addition of fifty beds, a new operating suite, a recovery room, a central supply service, emergency rooms, outpatient clinics, a physical therapy department, a pharmacy, a delivery room, nurseries, a linen room, a clinical laboratory, an X-ray suite, and a cardiopulmonary department.[260] Administrators were planning to add another twenty-four beds. An innovative dietary department, dining rooms, admitting offices, and gift and coffee shops were also expected shortly, while an updated power plant was in the design stages.

Saint Vincent's Hospital in West New Brighton celebrated the new Cardinal Spellman Pavilion in 1961. The pavilion had increased bed capacity to 310. But with the opening of the bridge, the Sisters of Charity knew there was more work to do. Expected future progress included expanding patient care, adding teaching programs, and increasing research. The structure also allowed for the first two-year nonresident school of nursing in the state, and the first premature center on Staten Island. Next up were residency programs in surgery, general medicine, pediatrics, and child psychiatry. A number of new full-time director positions would open, as would a modern intensive care unit. A department of psychiatry, a project with the Staten Island Mental Health Society, was also planned. Phase two of Saint Vincent's would include "extensive development of the radiology, emergency and clinic departments and installation of a new boiler plant."[261] Improvements for the X-ray department and emergency room were also being prepared. Saint Vincent's had grown from a modest fifteen beds when it opened in 1903 to being the "borough's largest general care institution" in 1965.[262]

In response to the borough's growing population, Staten Island Hospital called for an expansion circa 1963. Administrators stated that the number of general care hospital beds had not grown in correspondence with the rising population. In 1963, Staten Island Hospital admitted 10,357 individuals, an increase of 200 patients from 1962. The number of emergency room treatments also rose, as did laboratory exams. There was a slight decline in births.

The hospital was also planning for the future. This included the installation of electronic data processing machines in the accounting department. These machines would also provide medical statistics. This might seem provincial now, but it was an important step during the early 1960s. Supply inventory, patient records, and other data would soon be easier to analyze with this technology. A much-needed cancer detection center was opened in 1963, while the hospital had also joined the cancer tumor registry program of the American College of Surgeons. Both advancements would help cancer patients, the numbers of which were continually climbing. The Brewer drug dispensing system was introduced on all patient floors. This system made it easier to monitor drug accuracy, maintain records, and watch drug costs. A modern cardiology unit also appreciated by the community. Staten Island Hospital had expanded to include 251 beds by the fall of 1965. Further growth was planned.

In Clifton, the U.S. Public Health Hospital was operating for the benefit of government employees and merchant seamen. It was the oldest institution on Staten Island, having opened as the Seaman's Retreat in 1831. Sailors who entered New York Harbor were treated there for injuries and ailments through the nineteenth century and into the twentieth century. The site had many names over the years. Now operated by the federal government, it had 716 general care beds in 1965, but it served the entire metropolitan area.[263]

Sea View Hospital, opened in 1913 for tuberculosis patients, was nearly obsolete in 1961 owing to the cures developed for the deadly disease. By this time, Sea View and the Farm Colony had merged. It was now a sprawling four-hundred-acre campus. City officials and hospital administrators were seeking a new purpose for the site, so in 1964, a preliminary plan for an addition was announced. It was believed that the structure would accommodate three hundred people. In 1964, one thousand men and women resided at the site; all were elderly. Sea View Hospital and Home was and is the only municipal site on the island. Only sixteen of its 912 beds were for general care in 1965; all others were for the elderly. Staten Island is the only borough without a municipal general care hospital.

Preparation for the construction of a psychiatric hospital in Ocean Breeze was underway by the end of 1964. The State of New York hoped building would commence in early 1966. Initially, the hospital was to have three thousand beds, but with the opening of the bridge and improved access to the rest of the city, the number of planned beds was going to be updated. In order to prepare properly, the Department of Mental Hygiene was conducting a "future-growth survey" on Staten Island. At the conclusion of the study, the state would announce the size of the facility. The two-hundred-acre tract was bounded by Seaview Avenue, Seaside Boulevard (now Father Capodanno Boulevard), Mason Avenue, and Quintard Street.

No look at Staten Island during the 1960s would be complete without discussing the former Willowbrook State School on Victory Boulevard in the community of Willowbrook. The site opened in October 1947. The original purpose of the school was diverted for the care of injured military personnel during World War II. During the war, it was referred to as Halloran General Hospital. Afterward, it became Halloran Veterans Administration Hospital.

By 1951, 2,840 developmentally disabled patients were being cared for at Willowbrook State School. The following year, it was almost at capacity, so six more buildings were constructed to expand the program. Construction on only one structure had begun by March 1957. The next four buildings followed shortly thereafter. Further expansions led to it becoming the largest facility of its kind in 1960; as such, it was filled beyond its new capacity in 1961. A new institution for basic research on developmental disabilities was announced in 1964. It would be erected at Willowbrook.[264]

Willowbrook State School was severely overcrowded. There were more than six thousand patients at the facility in 1964. Physicians encouraged parents to hospitalize their developmentally disabled children at the facility, leading to a waiting list. To have a five-year-old child admitted took an average of eight months. Underfunded, understaffed, and lacking appropriate space to treat patients was a formula for a fiasco. Many of the patients were also blind, deaf, or had neuromuscular disorders. Training the staff of two thousand to work with the residents was a monumental task on its own. In addition to caring for the patient population, the hospital had another mission: to find the causes of and cures for disabilities.[265]

Willowbrook consisted of thirty-four buildings spread across four hundred acres of land. Its annual budget was $11 million. The property was bounded by Victory Boulevard, Forest Hill Road, and Willowbrook Road. Like other Staten Island facilities where large numbers of people lived together and were cared for by a supervising entity, it was a small city unto itself.

Willowbrook had its own police, a golf course, a pool, and its own generator in case borough power went out. Plumbers, bakers, roofers, gardeners, shoemakers, carpenters, and barbers were employed on-site, as were nurses, doctors, medical technicians, aides, and therapists.[266]

Residents ranged in age from infancy to seventy. They had various levels of disabilities. At this time, 1964, overcrowding was somewhat alleviated by transferring patients to other facilities. Another school that was under construction on Long Island in March 1964 would soon accept additional patients from Willowbrook.

Willowbrook offered academic courses, training programs, rehabilitation, occupational therapy, recreation, and religious programs for its residents. Approximately one thousand children attended these classes. Administrators were hoping to expand these programs. In addition, there were classes for children who were blind and/or deaf. Some six hundred adult patients worked in the industries located on the grounds. Women had laundry, kitchen, and dining duties, while men worked in the garage, the heating plant, and the industrial shops. More than one hundred patients held jobs in the local community.[267]

In the fall of 1971, it was learned that Willowbrook was not always what it appeared to be. Jane Kurtin of the *Staten Island Advance* wrote several in-depth stories about the facility. Her reporting uncovered numerous atrocities. The following year, Geraldo Rivera of the local ABC affiliate gained access to the school, accompanied by a cameraman, and the horrid conditions he saw were shared with the world. Meanwhile, parents of those in the school were vocal in seeking improvements. They were heard, and a class action lawsuit was established. It declared that residents had a constitutional right to be protected from harm.

> *In 1975, a Consent Judgement was entered in the 1972 Willowbrook lawsuit. That Judgement ordered that Willowbrook residents receive humane treatment and adequate clinical and educational services. This also set in motion the eventual closure of Willowbrook in 1987. It also began the development of community-based services. A mandated reduction from 6,000 to 250 residents was ordered by 1981.*[268]

As Staten Island's population grew, the construction of additional nursing homes was being considered, as was the expansion of existing nursing homes. Eger Nursing Home was preparing for the future. Administrators announced a 1966 expansion plan in celebration of the facility's fiftieth anniversary. A six-story, $3.4 million facility with 240 beds was announced. The building would have an unobstructed view of the surrounding woodlands. The nonprofit Eger Home was named after founder Carl Michael Eger, who opened the first home on Pulaski Street in Brooklyn in 1916. In 1926, seven senior citizens moved into a residence on the Egbertville property after the Brooklyn site was sold. A new structure, with thirty units, opened in 1957. It supplemented twenty units in an older frame residence that was still in use in 1966. The only other nursing home on Staten Island was Lily Pond Nursing Home in Fort Wadsworth. Silver Lake Nursing Home in Brighton Heights was in the planning stages. The first patient was admitted there on December 20, 1966. The three-story home cost $500,000 to build. It could accommodate 120 individuals, and it was expected to employ 45. It is located on Castleton Avenue.

Throughout the 1960s, the use of illegal drugs rose dramatically. While this was not an isolated problem on Staten Island, it did raise the crime rate of the borough. In addition, many residents and families experienced the heartache that accompanied the increased use of narcotics, particularly heroin. The availability of the Verrazzano-Narrows Bridge not only increased the flow of drugs but also allowed the borough to be seen as a location for addiction treatment centers. Many islanders were receptive to helping addicts; many were not. The beginning of drug addiction on Staten Island in that era may have begun with a short article in the *Staten Island Advance* on October 20, 1962. District Attorney John Braisted announced that there were thirty-five known narcotics addicts on Staten Island. According to the DA, most were teenagers. Braisted claimed there were "no real pushers of dope in the borough." He noted the relationship between addiction and crime and added that an officer with narcotics squad training had been placed in each of the three police precincts in the borough.[269]

Chapter 18

THE SHOPPING EXPERIENCE EVOLVES

Main Street shopping predominated into the late 1960s, but owing to Staten Island's population growth, more and more residents began shopping in the larger stores that opened during that decade. Sales tax savings in New Jersey encouraged off-island purchasing, as did the cheaper gas prices that were available in that state. The arrival of the bridge brought thirteen new shopping centers to Staten Island. It was hoped that these retail outlets would inspire shoppers to stay in the borough. The first major shopping center was located at the former Staten Island Airport on Richmond Avenue in New Springville. This airport was the last of four that once operated on Staten Island. The twenty-seven-acre shopping area that replaced it would be called Staten Island Downtown Center. This name was soon forgotten. When complete, there would be three hundred thousand square feet of shopping space. Parking for two thousand autos would be available. E.J. Korvettes would be the anchor store. A bank, an auto and tire repair shop, a supermarket, a furniture and carpet center, and other retailers would also open. A second phase of building would bring a movie theater, a variety store, specialty shops, and office space.

Thousands of shoppers descended on Korvettes for its official opening on September 24, 1964. It wasn't long before the parking lot was filled to capacity; those who arrived after that were forced to find street parking. After a brief ceremony, the public entered the store with a chorus of oohs and aahs.[270] Men's cashmere coats were thirty-nine dollars, while golf clubs could be purchased for sixteen dollars. Shoppers were mesmerized by an array of grand opening specials that kept cashiers tapping away at their machines. Not only were the bargains enticing, but the public also enjoyed wide aisles, air-conditioning, escalators, and a gorgeous white terrazzo floor. At the time, Korvettes had six hundred employees. Another four hundred would be hired in the upcoming months.[271]

Macy's New York and Sears, Roebuck and Company announced a joint purchase of 132 acres of New Springville land in October 1964. Adjacent to Korvettes, they would establish a modern mall. A third large retailer was expected (but did not arrive for many years), while sixty specialty stores would also be launched. The president of Macy's predicted that this retail center, which would eventually be called the Staten Island Mall, would employ around 3,500 individuals by 1970.[272] When the Staten Island Mall did open, it and the previously mentioned Forest Avenue Shoppers Town in Port Richmond Center (the Plaza) would cause the demise of many "Main Street" stores, especially along Richmond Avenue (now Port Richmond Avenue) in Port Richmond and Main Street in Tottenville and in the area around Tappen Park in Stapleton, as well as retail areas at New Brighton and West New Brighton. It would be many years before businesses returned to these locations. Only New Dorp Lane and Amboy Road in Great Kills would weather the arrival of the new concerns.

The retail food business saw an immediate impact on sales after the opening of the bridge. The number of mobile produce dealers, better known as fruit peddlers, dwindled to three in 1965. By this time, Charles Simon of New Brighton had been peddling fruit and vegetables for forty years, the past twenty-nine of them on Staten Island. His father taught him the business while he was selling bananas on the Lower East Side. Charles was given the overripe bananas, which he would sell to neighbors and friends. In 1965, he was using a 1959 International walk-in truck to move his produce, traveling as many as fifty miles a day. He could stay in business if he had regular customers. On the move six days a week, he was in Port Richmond on Mondays and Thursdays. Tuesdays and Fridays saw him vending around Castleton Corners and the South Shore, while on Wednesdays and Saturdays, he traveled around Grymes and Emerson Hills. At five thirty in the morning, he would arrive at the Staten Island Produce Company warehouse in Port Richmond to pick up produce. Most days concluded around six o'clock after he stored his leftover fruits and vegetables. Many hours of hard work allowed him to make a living. He did not want his sons following in his career choice, so he paid for some of their college educations.[273]

Chapter 19

RECREATION, STATEN ISLAND STYLE

Many island acres were under the domain of the New York City Parks Department. A proliferation of preserved land was located at Clove Lake Park, Silver Lake Park, LaTourette, and Wolfe's Pond Park. Expectations were high for additional parks. This was spurred on by the belief that Great Kills Park would be the "Jones Beach of Tomorrow." With Robert Moses at the helm, this was not surprising, as he was a prolific park builder. Most unfortunately, island beaches were polluted. Those who did venture into the water often came out coated in an oily film. Sewage and garbage added to the detritus. This viral cesspool of waste smelled and looked awful. The only nature preserve was the William T. Davis Wildlife refuge on Travis Avenue, at the border of New Springville and Travis. Originally, it was called the New Springville Bird Sanctuary.

As the borough's population rose, the Parks Department determined that staffing needed to increase. It was noted in 1964 that Parks Department facilities on the island had increased by 64 percent. The number of borough playgrounds was also scheduled to increase. In fact, these plans, in conjunction with Board of Education plans to establish three playgrounds adjacent to junior high schools and ten playgrounds adjacent to public elementary schools, would increase traditional parkland exponentially. Also in the planning stage was the War Memorial Ice Skating Rink at Clove Lakes Park. Land had also been acquired at Lemon Creek for the establishment of a park, as had land for an addition to Great Kills Park. Funding for additional acreage at both Wolfe's Pond Park and the Conference House Park was also being sought. Construction of a parking

The Billopp House, now known as the Conference House, in Tottenville as it appeared in the 1960s. *New York Public Library.*

lot and a children's zoo at the existing Staten Island Zoo were two other important projects in the works. A ball field at Burnside Avenue and Harrower Street in Westerleigh had also been created, but it would take years for the rest of Northerleigh Park to be improved.

There was an increased interest in improving what is now Historic Richmond Town from the mid-1950s into the 1960s. While the city did successfully accomplish a number of projects and enhancements, a hoped-for overhaul of the entire site was never realized, although a grand diorama of its completion was created. Early on, Robert Moses proposed that the Triborough Bridge and Tunnel Authority fund major improvements at the site. Unfortunately, the request to reroute the proposed Richmond Parkway through the Greenbelt by environmentalists piqued Moses's displeasure, so the promised money was stalled unless the original route was guaranteed. The situation became murky. Even so, the Lake-Tysen House, the cooper's shop, the basket maker's shop, and the Boehm-Frost House were all moved to the site during the 1960s. In 1969, a sawmill and the millpond area were completed. This was made possible by funding from the TBTA and the city. Also, the Third County Courthouse, the carriage house, and the basket maker's shop were opened to the public. Additional structures were designated New York City landmarks, while the former Colon General Store was moved from its South Shore location to the Restoration's campus. It was hoped that the Washington Hook and Ladder Firehouse, built circa 1890, would be moved from Charleston to the site. Plans also called for a Federal-period farmhouse in Greenridge, the Greek Revival Tysen House near Sailors' Snug Harbor, the Dr. Golder House in Rossville, and a Tottenville building known as the Tabernacle to be transported to the Restoration. The last structure served as an early 1800s Methodist meeting place. Moving these buildings would have cost $250,000.[274] They were never moved, but the historic circa-1720 Christopher House on Willowbrook Road, where Revolutionary War meetings and rendezvous took place, was reconstructed at Richmondtown between 1974 and 1978.

The fight to preserve the Greenbelt was a direct result of the opening of the Staten Island Expressway and its arteries. To prevent mass destruction, many environmentalists called for the Richmond Parkway to be altered or permanently removed from New York City maps. As of this writing, it survives on maps. Even so, parts of the Greenbelt have survived thanks to the Staten Island Greenbelt Natural Areas League (SIGNAL) and to Protectors of Pine Oak Woods. Today, the Greenbelt consists of High Rock Park, the William T. Davis Wildlife Refuge, Latourette Park, Deere Park, Willowbrook Park, Reed's Basket Willow Swamp, Bloodroot Valley, Buck's Hollow, Heyerdahl Hill, and a number of properties that do not come under the jurisdiction of the Parks Department. These include Seaview Hospital and the former Farm Colony, the Jewish Community Center, Pouch Boy Scout Camp, the Richmond County Country Club, Camp Kaufman, Moravian Cemetery, the College of Staten Island, and other contiguous lands. Today, the Greenbelt has a staffed nature center, a native plant propagation center, a carousel, a recreation center, a golf course, an archery field, ball fields and more.

Founded in 1881 as the Natural Science Association, the association, now known as the Staten Island Museum, has been at the forefront of environmental preservation since its founding. The founders were in fact collecting items from nature since the group's beginning. They soon noticed that the history of Staten Island needed to be preserved as, even in the 1880s, the island was undergoing vast changes. The founders of Protectors of Pine Oak Woods walked in the footsteps of these early naturalists and historians. They used their guiding influence as they preserved such locations as Clay Pit Ponds State Park Preserve, Blue Heron Park, Kingfisher Park, the Bluebelt, sections of the Greenbelt, and other unique areas.

In 1969, two thousand acres of national park land was proposed for New York Harbor. This would include land at Breezy Point and Jamaica Bay,

Queens; Floyd Bennett Field, Brooklyn; Great Kills, Hoffman Island, and Swinburne Island, Staten Island; and Sandy Hook, New Jersey. Eventually, the Staten Island unit would include Miller Field and Fort Wadsworth. The park would be created to meet the recreational needs of a growing urban public. It would be called Gateway National Recreation Area, and it would be administered by the National Park Service. Gateway National Recreation Area had the backing of President Richard Nixon. Miller Field in New Dorp was deactivated by the U.S. Army in 1969. Canarsie Pier, Fort Tilden, and Jacob Riis Park are now also a part of Gateway National Recreation Area. Green spaces, beaches, historic structures, and cultural landscapes can all be enjoyed.

The Fresh Kills Landfill was busy around the clock in the 1960s. Staten Island residents were always told it would become a park. Many laughed at the idea, but the former dump is now being transformed from the largest dump in the world to the largest park in New York City. It currently includes 2,200 acres. Recent activities have included kayaking workshops, a Raptor Fest, photography classes, bird-watching, wetland study, and assorted activities that focus on ecology, science, and art. Plans for ball fields, a boat launch, art installations, and horseback riding trails are being developed. It will be an important site for educational studies and athletic activities as the population of Staten Island continues to expand.

Historically, movie theaters have been one of the most popular venues for entertainment. During the 1960s, a number of historic theaters operated on the island, including the Ritz and the Empire in Port Richmond, the Stadium in Tottenville, the Lane in New Dorp, the Liberty and Paramount in Stapleton, and the Saint George Theatre in Saint George. A drive-in theater on Richmond Avenue in New Springville was also a well-liked attraction, but it eventually closed; the land was much too valuable.

In New Springville, two movie theaters opened to cater to the existing and recently arriving Brooklyn population. A company called Century Theaters built a 756-seat theater on the east side of the Korvettes Shopping Center. It opened on June 14, 1967. The company even held a contest to name this new showplace; 250 names were proposed. Springville, Centurion, and Verrazano were among those recommended. Eventually, the theater was called Century's Richmond Theater.

Opening night on June 14, 1967, was a festive event. The Emeraldettes, a sixty-girl drum and bugle corps of the Tottenville Precinct Police Athletic League, provided loud and lively tunes. They were joined by an array of colorful flags that floated on the breeze. Floodlights showcased the event.

The opening was also a benefit for the Great Kills Community Center. The organization was hoping to build a new site. Also in attendance was Frank Roma, an island recording artist, who sang to the music of Bob Soldivera and his orchestra. One interesting feature of this modern theater was that a ten-foot-wide canopy over the box office could be extended to thirty-two feet to protect patrons when it rained. Opening night featured *Casino Royale*, the latest James Bond film. Bond was played by David Niven.

Another theater opened on Richmond Avenue in New Springville on June 28, 1966. Known as the Island, it had one thousand seats. It was owned by the Fabian Theater chain, which also owned the Saint George, Paramount, and Ritz theaters. *My Fair Lady*, three hours in length, was the featured film. The evening was dedicated as a benefit for the Staten Island Chapter of the American Red Cross. Attired in cocktail dresses, spike heels, furs, and theater suits, one thousand theatergoers attended. Outside, two Hollywood-style searchlights danced across the misty sky.

The Fox Plaza Theater, located at 105 Mill Road in New Dorp, opened on April 7, 1966. Legendary actress Joan Crawford arrived with a troupe of Disney characters featuring Winnie the Pooh and his colleagues. Guests dressed in gowns and tuxedos greeted Crawford. After the ribbon cutting, all were treated to a doubleheader featuring *The Ugly Dachshund* and *Winnie the Pooh*. The event was also a fundraiser for the local branch of the American Cancer Society. An *Advance* reporter noted that the opening of both the Fox Theater and the Island in New Springville signaled the end of a dark era for Staten Island theaters. He noted that it had been decades since a new theater had opened. The reporter observed that many island movie theaters had closed or been repurposed, including the Liberty, Rex, Park, and Richmond theaters in Stapleton. The Palace in Port Richmond, the Plaza and Capitol in West New Brighton, the Stadium in Tottenville, the Star in New Brighton, the Staten in New Dorp, the Strand in Great Kills, the Victory in Tompkinsville, and the drive-in had all ceased operation.

Entertainment and culture again interacted when *Sleeping Beauty* as performed on film by the Kirov Ballet was shown on May 4 and 5, 1966, at the Fox Theater. The Leningrad Kirov Ballet was said to have the "oldest continuous tradition in the dance world. Its antecedents go back to Imperial Russia, actually to Catherine the Great [1729–1796] and her son Paul I, who sponsored development of the Imperial Russian Ballet." The following night, ticket holders watched the always popular film *The Trouble with Angels* starring Rosalind Russell and Haley Mills. It was advertised as a "hip, hilarious, heart-warming honey of a movie."

Chapter 20

PRESERVING THE HISTORIC PAST

The unnecessary teardowns of historic structures that occurred in New York City during the 1960s were destroying the built environment of the five boroughs. The first reaction came from Manhattan residents. In response, the Landmarks Preservation Commission was established in 1965. The commission was tasked with responding to the concerns of New Yorkers who feared that important physical elements of the city's history were being lost to demolition despite the fact that many of the buildings could have been reused instead of destroyed. Events like the demolition of the architecturally distinguished Pennsylvania Station in 1963 spurred the movement onward.

On Staten Island, there were numerous examples of historic destruction, including one building that is still mourned today. The so-called Horrmann Castle came to a crashing end on January 23, 1968. The mansion was built by beer baron William Horrmann, son of August Horrmann, cofounder of the Rubsam & Horrmann Atlantic Brewery, in 1910. It was described as the most fanciful structure ever built on Staten Island. Designed in the fashion of castles overlooking the Rhine River in Germany, it had forty rooms, a 440-foot tower, a ballroom, and a solarium. The landscaped grounds consisted of three terraced levels.

In 1945, the Presentation Sisters bought it for use as a motherhouse—or, as it is better known, a convent. By 1965, the "castle" was no longer a viable residence for the Sisters, so they sold it to real estate developers. The massive mansion crashed to the ground after a ninety-minute battle with a bulldozer. This is just one of several magnificent buildings that were razed during the building boom resulting from the construction of the Verrazzano-Narrows Bridge.

Our Lady of Pity Roman Catholic Church stood at Lamberts Lane and Richmond Avenue in Graniteville. It was demolished during the summer of 1960. It stood in the path of the Staten Island Expressway. Our Lady of Pity Church had humble beginnings. Graniteville parishioners began meeting in private homes to celebrate mass in 1913. The parish was an off-shoot of Our Lady of Mount Carmel in West New Brighton. In 1919, a prefabricated church was built on Sumner Avenue. Not only did the local inhabitants and farmers from Graniteville attend, but nearby residents of Bulls Head and Willowbrook also joined the congregation. Owing to the numerous Italian immigrants moving to the area, the mission became an Italian National Parish in 1923, and a traditional church of that era was built. When this church was condemned, Monsignor Anthony Ricotti built a parish hall on previously purchased land at 1616 Richmond Avenue in Bulls Head. It was here that masses were held until a modern church opened in 1988. Both structures still exist today. In 2015, Our Lady of Pity merged with Saint Anthony of Padua in Travis.[275] The expressway's path also signaled the demise of Saint Simon's Episcopal Church in Concord. It was demolished in the fall of 1959. By the following year, a new church stood on Richmond Road in Dongan.

The expressway's path signaled the demise of Saint Simon's Episcopal Church in Concord. It was demolished in the fall of 1959. *New York Public Library.*

Public School 34 in Fort Wadsworth was another institution destroyed for construction of the expressway. In fact, to build the highway, 400 buildings were demolished overall, while 3,500 residents were displaced in Fort Wadsworth, Concord, Sunnyside, Willowbrook, Graniteville, Old Place, and elsewhere.[276]

Other historically significant structures that were destroyed due to the construction of houses, roadways, stores, and apartment buildings on Staten Island after the opening of the Verrazzano-Narrows Bridge included a glorious seventeenth-century building with a spring eave above a two-story portico that had octagonal columns. This building stood at 2141 Richmond Avenue at the corner of Rockland Avenue in New Springville. At one time, it was known as the Darcey Hotel. In 1922, the building was adorned with various signs advertising Castle's ice cream, Michelin, and the fact that two telephones were available for the public inside the building. Around 1920, the hotel was owned by Gerald Darcey. Around 1965, the house was demolished, even though the owner offered it for sale to any individual who would pay one dollar for the structure. The stipulation was that it would have to be relocated to prevent demolition.[277]

Seeking to add usable space for its increasing number of congregants, the Woodrow Methodist Episcopal Church on Woodrow Road had the Captain Abraham Cole house razed in 1967. A parish hall would be built on the site. Cole's grandfather, also Abraham Cole, had been instrumental in donating money for the construction of the original church in 1787, and he and his ancestors had spent countless hours at this house of worship. The house was said to feature one of the finest examples of a Colonial curved roof portico with Greek Revival columns. Even the Landmarks Preservation Commission included it on a list of Staten Island structures that should be considered for landmark status. The proposed demolition of the Captain Abraham Cole house, believed to have been built around 1836, was roundly criticized, but its destruction went on as planned.

Even with the obliteration of historic buildings on Staten Island, many structures were landmarked. They included Battery Weed (Fort Wadsworth, landmarked in 1967), the New Dorp Lighthouse (landmarked in 1967), and the Charles Kreischer Mansion (Charleston, landmarked in 1968). The Conference House in Tottenville was landmarked in 1967. It was at this site on September 11, 1776, that a failed attempt for peace during the Revolutionary War was made when Patriots John Adams, Benjamin Franklin, and Edward Rutledge met with Lord Admiral Richard Howe of the British Royal Navy.

A phenomenal number of structures associated with the Richmondtown Restoration (now Historic Richmond Town) were landmarked in the 1960s and the 1970s. These included:

- The Voorleezer's House (circa mid-1700s, designated 1969)
- Treasure House (circa 1700 with additions, designated 1969)
- Rezeau-Van Pelt Homestead Graveyard (established 1700s, designated 1969)
- Christopher House (circa 1720 with a later addition, designated 1969)
- Guyon-Lake-Tysen House (circa 1740 with later additions, designated 1969)
- Boehm House (1750 with a later addition, designated 1969)
- Kruser-Finley House (circa 1790 with additions, designated 1969)
- Basket Maker's House (circa 1810–1820, designated 1969)
- Sylvanus Decker Farm, 435 Richmond Hill Road (circa 1810, designated 1969)
- Third County Courthouse (1837, designated 1969)
- Stephens-Black House (circa 1838–1840, designated 1969)
- Bennett House (1839 with a later addition, designated 1969)
- Richmond County Clerk and Surrogate's Office (1848 with later additions, designated 1969)
- Parsonage (1855, designated 1969)
- Eltingville Store (circa 1860, designated 1969)
- Britton Cottage (circa 1670 with later additions, designated 1976)
- Billiou-Stillwell-Perine House, 1476 Richmond Road, New Dorp (circa 1663 with later additions, designated 1967)

Buildings from the previous three centuries were threatened from one end of Staten Island to the other. For this reason and others, several historic buildings were moved to Richmondtown. They were:

- Boehm House (circa 1750, moved from Great Kills on June 23, 1965)
- Finley House (now known as the Kruser-Finley House, circa 1790 with additions circa 1820 and circa 1840–1890, moved from Richmond Road in Egbertville in 1965)
- New Dorp Railroad Station (1889, moved around November 2, 1965)
- Britton Cottage (moved from New Dorp around October 29, 1965—and relocated none too soon, as someone had attempted to burn it the previous year)

The Billiou-Stillwell-Perine House at 1476 Richmond Road is owned and maintained by the Staten Island Historical Society/Historic Richmond Town. *Undated photo, New York Public Library.*

The Britton Cottage at its original location on New Dorp Lane in New Dorp, circa 1907. *New York Public Library.*

The foresightedness of several individuals supported the preservation of numerous buildings and sites. In the 1960s, these individuals included Loring McMillen, Marjorie Decker-Johnson, Brad Greene, Dorothy Valentine-Smith, Mathilde Weingartner, Gail Schneider, Carlton and Louise Beil, Howard Cleaves, James Whitford, Raymond Fingado, Terence Benbow, Robert Hagenhofer, Marjorie Kerr, George Pratt, and others.

The tail end of the 1960s saw a tremendous effort by preservationists to save the Alice Austen House in Rosebank. There were plans to demolish the house, which was in a state of disrepair, so an apartment building could be built. It would overlook New York Harbor and the Narrows. Renters would have a front-seat view of the Verrazzano-Narrows Bridge. The house had been the home of photographer Alice Austen and her partner Gertrude Tate. Alice's grandfather John Haggerty Austen bought the 1690 cottage in 1844. It was to be a summer getaway for this Manhattan family. Austen took to repairing and enhancing the structure with the result that the family moved there year-round in 1852. They named it Clear Comfort. A picturesque suburban cottage, it is very similar to other structures that were built along the shore during the nineteenth century. Today, the cottage is an historic house and museum that celebrates the life of Alice Austen, a woman who took thousands of images that documented not only the well-to-do but also the everyday lives of those who lived and labored at Staten Island, Manhattan, and elsewhere. Austen's documentation allows researchers, historians, and the public to see into our historic past. Not only is the home a New York City landmark; it is also on the New York State and National Registers of Historic Places. In 2017, it became a designated National Site of LGBTQ+ History.

Another threat that came from the opening of the Verrazzano-Narrows Bridge and the corresponding highways that opened with the span was the Shore Front Drive. The Austen House stood in the middle of a proposed route for the northerly extension of this road. Fortunately, the Austen House was not destroyed.

One aspect of historic preservation that came to the forefront owing to the residential overbuilding that resulted from the construction of the Verrazzano-Narrows Bridge was archaeology. Scholars were racing to uncover artifacts before they were covered over by bulldozers. Digs were speedily undertaken by New York University, Columbia University, the Staten Island Historical Society, the Staten Island Institute of Arts and Sciences, the New York City Parks Department, the Brooklyn Children's Museum, the Museum of the American Indian, and others. They steadfastly worked to set up a coordinated program of archaeological research on the island, even before the bridge opened, in order to safeguard the history of the Lenape and the island's prehistoric populations.

Staten Island's most important archaeological site is located at Wards Point in Tottenville. Around 1960, Dr. Jerome Jacobson made a thorough study of the site. He also compiled information on professional, semiprofessional, and amateur digs that took place there over the previous one hundred plus years. His findings were published in 1980. The report *Burial Ridge: Tottenville, Staten Island, New York: Archaeology at New York City's Largest Prehistoric Cemetery* is the most well-regarded publication on this site. In 1858, an entire skeleton and several skulls were discovered by some very surprised workers who were digging a house foundation for a man named Joel Cole. In 1982, the Ward's Point Conservation Area, including Burial Ridge, was listed on the New York State and National Registers of Historic Places. It was noted that the area was active from pre-Columbian times until the arrival of Europeans.

Chapter 21

STATEN ISLAND RESPONDS

The number of civic associations in Staten Island jumped during the mid-1960s. This was mainly in response to problems with new construction. One group that formed around 1964 was the Duncan Ridge Civic Association. It sought to persuade a builder to complete paving Duncan Street between Rome and Sparkhill Avenues in Dongan Hills. Only one side of Duncan Street was paved when construction was completed. It seems that the builder who built odd-numbered houses paved his portion of the street, and the even-numbered house builder did not. The association was successful in having the city remedy the problem. Once that effort was complete, the civic group turned its attention to other matters, including beautifying the neighborhood, blocking the construction of a garden apartment and the Richmond Parkway, and air pollution.[278]

One group that formed under the umbrella of the Prince's Bay-Pleasant Plains Civic Association was the Committee for the Preservation of Amboy Road. Between one hundred and two hundred homeowners were interested in New York City landmark status for a housing district to protect their community. Amboy Road held many fine historic structures, and residents were aware they needed to be safeguarded and preserved. Unfortunately, this did not occur. Had such an effort been successful, many more historic showplaces would still be standing on this roadway.

The Decker Avenue Civic Association called for the Department of Marine and Aviation to cleanup Staten Island's ferryboats. The organization believed that if workers could not complete the task, a private company

Barrett Park or Staten Island Zoo, West New Brighton, circa 1940. *New York Public Library.*

should be hired. During the mid-1960s, the group covered such topics as fluoridation, free college tuition, school naming, and pollution.

In 1965, Jim Scarcella was president of the Forest Heights Civic Association. It focused on the triangle formed by Forest Avenue, Broadway, and Clove Road in West New Brighton. Owing to Clove Lake Park, the Staten Island Zoo, and the YMCA, the area was a traffic nightmare around 1961. The association fought to have the area south of the zoo between Broadway and Clove Road turned into a series of one-way streets. It also changed the location of the children's zoo before it was established. The association feared that animals would be placed close to residences. By 1965, it was attempting to create a positive community spirit in spite of all the changes.

The Chestnut Hill Civic Association was busy battling a pipeline for aviation fuel in 1965. It was to run from Linden, New Jersey, across Staten Island to Kennedy Airport. It would move five thousand barrels of fuel per hour. The organization feared that explosions would occur owing to leaks. Most group members were residents of Todt Hill. They wanted the pipeline away from their homes and installed alongside the Staten Island Expressway right of way.[279]

In the summer of 1965, a group of mothers demonstrated against a pipeline. They hoisted placards reading "What will the pipeline do for us?"

and "Save our street: Safety first" as laborers laid fuel lines on Saint John's Avenue in Rosebank. The hour-long protest featured approximately twenty moms and fifty children. Homeowners wanted the pipe buried twenty feet down, but the pipeline company would agree only to install the pipe seven feet down as opposed to the original six feet that was proposed. It was noted that the company was only required to bury the pipe three feet below the ground. It appears that the company installing the pipeline and the neighborhood got off to a bad start as locals were not informed that the street would be ripped up for the project. This was a borough-wide phenomenon that occurred frequently.

A short *Advance* article on January 18, 1967, informed the public that construction of a liquified natural gas (LNG) tank in Bloomfield would begin in the spring. The tank was to be built by the Texas Eastern Transmission Corporation and would cost $13.1 million. The concrete tank would stand on sixty-six acres west of Bloomfield Avenue. This was a little-known and little-used area unless someone wanted to illegally dump broken washing machines, refrigerators, or the like. The property had been purchased from the Gulf Refining Company around 1965. Two billion gallons of LNG could be stored in the tank. Per Texas Eastern, it made more sense to store gas where the consumers were located instead of continually piping it in. Gas would be distributed to Brooklyn Union Gas Company, New Jersey Natural Gas Company, and others.[280]

The tank went into operation during April 1970 when 240,000 barrels of LNG arrived from Algeria. Company officials insisted the tanks and associated pipes were safe. In fact, the New York City Fire Department had "originally denied the Texas Eastern Transmission Company a permit to build the tank" in Bloomfield…[but] was overruled by the city's Board of Standards and Appeals when the company agreed to institute extra fire-safety measures."[281] The fire department was wise to disallow the permit. On February 10, 1973, the tank exploded, and a fire blazed in its wake. Forty men were killed. They did not stand a chance as the explosion blew the concrete roof into the air. It fell back onto the men in the tank as they were cleaning it.

> *The dome-shaped tank had an aluminum liner insulated with polyurethane foam covered with mylar plastic lining. The plastic lining had been repaired with heat sealing, and an official theorized that heat, along with natural gas, had been stored in the foam, building up to the combustion point. A spark was believed to have touched off the explosion.*[282]

This was the first time this type of lining and insulation were used together in the United States to "keep the liquefied natural gas stored at 260 degrees below zero."[283]

Seven days after the blast, District Attorney John Braisted Jr. implored Mayor John Lindsay to immediately halt construction of two liquefied natural gas tanks along the Arthur Kill River in Rossville. Braisted insisted that "errors in construction or in provisions for safety, which may have been made at the Bloomfield installation, should not be repeated at Rossville."[284] The Rossville tanks were never filled. They have stood in a solemn state of limbo for five decades thanks to the hard work of a community organization called BLAST, the organization founded to prevent use of these gargantuan metal reservoirs. In the end, it was determined that they should finally be dismantled. In 2022, plans were aired to use the site as part of an offshore wind hub. As of this writing, the tanks are still standing.

In December 1966, the Clove Valley Community Association was battling to prevent the Fountains from building three more apartment buildings near the two that already existed on Clove Road adjacent to Clove Lakes Park in Sunnyside. The group was composed of twenty-eight local families who were alarmed at the potential density of a five-building complex. Lack of parking and the impact on the neighborhood's quality of life were major concerns. In fact, a minimal amount of work had already been done for the construction of buildings three and four. The group was successful. The two original apartment buildings are the only ones standing today.

The residents of recently built homes were also organizing. In 1971, the Heartland Village Civic Association banded together over a lack of fire hydrants and the fact that only two temporary alarm boxes were located in the community. One hydrant was buried up to its spigots. According to one report, it was useless to both firefighters and dogs. The association was equally concerned because the boilers in several recently constructed houses had caught fire. For these reasons, the builder and a fire chief attended one of the group's meetings to discuss the situation. The chief noted that it would be nearly impossible for the fire department to fight any inferno along Rockland Avenue, Forest Hill Road, Richmond Hill Road, and Richmond Avenue. Road conditions were so primitive that they would not support the twenty tons of equipment needed to reach a fire. The chief suggested that the builder install access roads. The builder responded that he would build roads on the land he owned but that the city needed to improve the overall New Springville approaches. All agreed that the planned widening of Richmond Avenue would help alleviate the problem. It was further noted that five permanent alarm boxes would soon be installed. Regarding the partially covered fire hydrant, the chief said the builder installed sidewalks that were too high. The builder claimed the city put in fire hydrants that were too low.[285] Shortly after this information appeared in the *Advance*, Richmond Avenue went from a winding country lane to an eight-lane speedway that was calmed only after dozens of traffic lights were installed.

The Staten Island Civic Congress announced its support of a city jail in the borough on December 11, 1964, but only if the prisoners were island residents. They also expressed gratitude to the City Planning Commission for its support of the proposed Greenbelt plan. The group went on to display an active interest in air pollution, rent control, and tax returns. The Civic Congress was comprised of several civic organizations that banded together.

Other civic groups that were active in the 1960s included the 10309 Civic Association, Annadale Community Association, Arlington Civic Association, Atlantic Avenue Association, Bay Terrace Civic Association, Butler Manor Civic Association, Castleton Corners Civic Association, Concord Civic Association, Eltingville Civic Association, Emerson Hill Association, Great Kills Civic Association, Livingston Community Association, Manor Civic Association, Midland Beach Progressive Association, New Dorp Civic Association, Oakwood Civic Association, Old Town Civic Association, Richmondtown-Clarke Avenue Civic Association, Rossville Civic Association, Silver Lake Civic Association, Sunnyside Community Association, Todt Hill Civic Association, Ward-Nixon Civic Association, Westerleigh Improvement Society, and the Woodacres Civic Association. This is not a complete list. It should be noted that the Westerleigh Improvement Society was established in 1887. It has existed for more than 138 years. Over the decades, the group has worked to preserve the historic qualities of the community through rezoning, public programs, publications, and more. The organization monitors everything from crime to sanitation to transportation needs. It is an incredible group that should serve as a model of excellence for those working to better Staten Island.

As noted, the parent–teacher associations of Staten Island were relentless in attempting to improve the quality of schools and education in the wake of the arrival of thousands of additional students before and after the opening of the Verrazzano-Narrows Bridge. In this situation, both natives and recent arrivals came together to voice concern and sometimes overturn problematic projects and ideas that continually surfaced as the

1960s unfolded. Civic groups formed from the PTAs whose members and administrators faced off with the City of New York, builders, and even corporations in an attempt to do what was best for schools and neighborhoods. Some PTA groups had been organized decades before the bridge opened; others came together as schools were opened.

Chapter 22

UNIFORMED PERSONNEL, UTILITIES, AND OTHER SERVICES

Given Staten Island's expected population increase, water experts urged the city to make progress on installing a water tunnel under the Narrows. It would run from Brooklyn to the island. In 1960, there were only two tiny siphons under the Narrows. Together, they delivered thirty million gallons of water a day to the borough. It was estimated that well water was providing five million gallons of water per day to some island residents. A new aqueduct with a ten-foot diameter would allow three hundred million gallons to reach Staten Island daily. The report also recommended installing covered tanks at Silver Lake Park. These tanks would hold one hundred million gallons of water. The deadline for the tunnel and the tanks was 1964—the projected year of the bridge's opening.

By the 1950s, Staten Island was the only borough without a direct supply of water from the city's reservoir system. Work officially began on the new aqueduct on February 20, 1962. On that day, Borough President Maniscalco and Deputy Mayor Edward Cavanagh Jr. ceremoniously set off a smoke bomb in the Tompkinsville shaft of the project near Pier 7 on the East Shore. It was expected to cost $25 million. Work had actually begun a few weeks before, but this event marked the official kickoff. Maniscalco stated that the project "staggers the imagination" owing to the breadth and scope of the work.[286] The new aqueduct was part of an even larger project of moving water from the Delaware River water system in the Catskills. Those who fought the plan suggested that water be obtained from the Hudson River. They suggested that if it was mixed with water from the existing Catskill and Croton lines, no one would notice the difference. The city refused, so the Delaware system was developed. This was one area where the City of New York anticipated the needs of the people of Staten Island before the bridge opened.

Unfortunately, when the project was completed, it was long after the fires of 1963. Installing a modern water tunnel correlated with improving firefighting on the island, especially with the opening of the bridge and the expected increase in population and structures. The need for additional water was burned into the souls of island residents on April 20, 1963, when three disastrous fires burned simultaneously at Tottenville, Rossville, and Mariners Harbor. That date is now known as Black Saturday. It was the day that more than one hundred homes were consumed by fire. In the Rossville area, it was said, every other house burned to its foundation. Damage was estimated at $2 million. Hardest hit was the historic African American community of Sandy Ground. The devastation was so staggering that even the firemen cried as homes turned to ashes.

Why was there so much destruction? The main problem for firemen was that water pressure was frightfully low, especially on the South Shore. Instead of gushing water, hoses literally dripped droplets. At the peak of the fires, pressure dropped to ten pounds per square inch. Forty to fifty pounds per square inch was considered normal. Citizens made bucket brigades in an attempt to squelch the flames or wet threatened houses. Some used sand and dirt to put out fires since there was so little water available. Firemen were further frustrated. As soon as one fire was put out, another sprung up. A dry spring, leaf litter, and wind created a recipe for catastrophe. Houses, cars, a gas station, telephone poles, woodlands, and even wildlife were incinerated.

Fire engines in Tottenville were on every street leading to the Raritan Bay. Saint Louis Academy harbored fifty women and children who received juice and cookies for nutrition and comfort. Initially it was denied that the lack of fire hydrants and low water pressure led to the catastrophe. Many complained that the number of fires put a strain on the water system. Others noted that firemen could not reach the hydrants because of overgrown brush. In reality, the water tunnel under the Narrows would improve water pressure and firefighting, but by 1963, it was behind schedule.

With continuous construction in the borough, the fire department saw fewer brush fires as woods and meadows were leveled to accommodate the 1960s building fury. But this meant that house fires would increase. Therefore, the department announced plans to build modern firehouses at Travis, Huguenot, and Eltingville. A modern fire center had also opened on Slosson Avenue in Castleton Corners in 1962. Designed to improve fire department coverage for Staten Island, it was also established to protect the city in the event of a nuclear attack, an ever-present fear during the Cold War. It cost almost $1 million. The builder claimed it would withstand a nuclear barrage unless it was a direct hit or near miss. It replaced a 1907 fire center that was located at Borough Hall in Saint George.

Part of Staten Island's appeal was always its connection to the past. One aspect of this was the presence of volunteer firefighting companies on the island. Two were still active in the 1960s and remain so today. The *New York Times* reported on May 7, 1960, that the Richmond Engine Company on Richmond Road in Richmondtown, founded in 1903, was preparing to close. The twenty-six men who then comprised the company decided to end the service since a New York City firehouse was scheduled to open on Richmond Road in New Dorp. Seventy-five-year-old Captain Leroy Barton, who was described as a "stubby, peppery electrician," was in charge of the house, which had one 1937 single-hose wagon. Each year, the city gave the company $1,200 to maintain the wagon. A potbellied stove kept the firehouse cozy. The oldest serving member was eighty-four, while the youngest was twenty-nine. On average, they answered approximately 125 calls a year. The firefighters mainly combatted the notorious brush fires of the borough. While a reporter interviewed the volunteers, a fire call was received. The men quickly responded. It was a brush fire, and the men extinguished it with Indian pumps. Staten Island's other volunteer fire company is the Oceanic Hook and Ladder Company in Travis, founded in 1880.

The water tunnel was behind schedule, but its construction continued. On December 1, 1966, 650 pounds of dynamite blew approximately one thousand feet under the Upper Bay to finish the link between Brooklyn and Staten Island. But work on the tunnel was not complete. It needed to be lined with concrete, which would take another year or more. Borough President Maniscalco also noted that many of the distribution lines on the island were too small to accommodate the additional water that would flow to the borough.

The tunnel contractor was tasked with doing something with the excavated material that was removed. Much of the rock was given to other companies, who, in turn, used it to fill wetlands—an appalling thought for today's environmentally conscious population. It was reported that much of the rock was dumped near the approach road of the Goethals Bridge. Other excavated material still sits around the Goethals Bridge Pond. Material was also taken to Eibs Pond in Concord and the shoreline at the base of

Westervelt Avenue in New Brighton, where it extended the waterfront. Some of the fill was dumped near Harbor Road and Richmond Terrace in Mariner's Harbor, where, again, the shoreline was extended.[287]

Setup for the installation of steel tanks at Silver Lake Park began in October 1967. Many of the park's neighbors and other islanders were concerned that the park would be destroyed by the undertaking. They had good reason to be. Preliminary plans called for the removal of one thousand trees. An eight-foot, half-mile-long wooden fence was preventing islanders from viewing the park. The lake itself was surrounded by wire fencing, so there was no need for the wooden fence. Pedestrians were also prohibited from entering the park; the contractor advised that it was too dangerous. These facts led the public to question the secrecy surrounding the project.

In order to install the tanks, eight hundred thousand cubic feet of earth needed to be moved. The soil would stay on-site and be utilized as the project progressed. A two-ton machine was ready to go by October 29, 1967. The not-so-magnificent monster could knock down trees at a rate of two or three acres a day.

The tanks would replace the existing open-air reservoir. The enclosed tanks would provide cleaner water as one source of contamination had been seagull droppings. Because the tanks would be sixty feet higher than the original basins and improved pipelines leading to the tanks would be connected, pressure would be improved. This was sorely needed on the island given that the lack of water pressure caused hydrants to run dry during the catastrophic fires of April 1963. Initially, it was believed the installation would take thirty months. In no time, project managers suggested completion would occur in 1971. Court action by local residents who were appalled at the destruction of the park also caused delays. This battle led to the formation of the Silver Lake Civic Association.

A structure to house the reservoir controls would be placed near the center of the tanks. It is the building that stands parallel to Victory Boulevard. The basins are part of the tank's draining system. When installed, the tanks were the largest of their kind in the world.

By April 1970, the water tunnel had been completed and work on the water tanks at Silver Lake Park was well underway. By late April 1972, the water tanks were in place. Only the landscaping needed completion. The entire project was finally dedicated on June 17, 1974. A bronze plaque to commemorate the project was unveiled. Two recent events did bring sadness. Two young men, George Amoia and Earl Sasso Jr., had recently drowned at the reservoir. Their friends and family were present, as was a large floral

Silver Lake Park was the site for two borough water tanks. They would improve water quality and water pressure. *New York Public Library.*

display dedicated to their memory. Borough President Conner walked around the two basins and afterward promised to install safety measures to prevent such events.

New York Telephone Company reported in 1960 that it was preparing for the influx of new residents. District manager Raymond Fingado noted there were 85,350 telephones on the island that year. Fingado predicted that in 1964, there would be more than 100,000 phones. He believed that "tasteful, better designed products" needed to be manufactured for consumers. As such, the new "Princess" telephone would be introduced. Called an improved phone, it was lightweight, had a nightlight, and came in five colors, including a modern shade of turquoise. New York Telephone also introduced the "Home Interphone" in 1960. It would enable residents to answer their doors by telephone while also allowing for room-to-room conversation. The future was fast approaching as testing was taking place to see if handsets with dials were functional. Farsighted staff were also looking into replacing dials with push buttons. Expected future products included phones compact enough to carry in a shirt pocket. These phones would allow the owner to make and receive calls while walking down the street. The forecast also included satellites beaming live television shows around the world.[288]

The installation of telephone lines boomed as the number of residents increased in anticipation of the opening of the bridge. Modern telephone cable was placed from one end of the borough to the other, while additional equipment was purchased for the New York Telephone company building in New Dorp in 1964. $1 million had been spent to upgrade the Staten Island system. Much of the upgrade required digging up roads recently laid to place additional cable. In that year, a record number of phone calls was placed; the average business day saw 470,000 daily.

The numbers continued to grow. Raymond Fingado stated that growth and modernization projects for the borough would total $3 million in 1966. This included building a one-story addition to New York Telephone's central office at 355 Forest Avenue in West New Brighton. A new switching center would be established here to meet the demand for communication.

The year 1969 would be another year of expansion for New York Telephone, Con Edison, and Brooklyn Union Gas. In February, 141,000 telephones were in operation, while miles of telephone line were strung and buried throughout the borough. Interestingly and historically, the first "computer-operated centrex telephone system appeared on the Island at Wagner College" that year.[289] High-speed and convenient, it allowed college staff to receive calls directly from the two hundred extensions that were installed across the campus. Borough Hall was next in line to receive a computer-operated phone system.

Rear view of the Staten Island Borough Hall, Saint George, circa 1920s. *New York Public Library.*

Con Ed had begun a five-year construction project in 1959. The company hoped to keep pace with an anticipated load growth both in the borough and beyond. The budget included demolishing the former Livingston generating station on Richmond Terrace after it was taken offline in late 1961. It was during the summer of 1959 that a 350,000-kilowatt generating station opened in Travis. A turbo generator, an up-to-date boiler, a coal-handling facility, and other site improvements were also planned for the Arthur Kill River location.[290]

By 1965, the Travis plant was producing most of the borough's electricity but not all of it. On Staten Island, Con Edison employed 486 individuals and serviced 77,410 borough customers.[291] Con Ed's growth would continue as the population increased. A plan to add a 500,000-kilowatt generating unit to the Travis plant was announced after the New Year in 1966. The contemporary unit would cost $68 million to build. According to Con Ed, it would be "equipped with air pollution control devices which…[would] remove 99.5% of the solids contained in exhaust gases." It was scheduled to open in 1968. Coal would also fuel this plant.[292]

By 1964, not only was air pollution hazardous, but it became even more so when combined with fog. Experts contended that the high sulfur content from fuel, including coal and oil, was the culprit. At this time, Con Edison was the largest fuel consumer in New York City. Representatives of the gas industry suggested using gas as an alternative to pollution-causing fuels. Con Ed was burning 2,600 tons of bituminous coal every day at its Travis plant. In 1965, coal was vital to producing electricity for Staten Island as it created steam for the turbines to produce electricity. The facility did have the capacity to also use oil.

By late 1967, Con Edison was preparing to enter the field of nuclear power by introducing nuclear power plants in the metro area. Initially, the island had been excluded from the plan, but in September that year, Con Ed called for building a nuclear power plant on either Hoffmann or Swinburne Islands off South Beach. This suggestion was ruled out when Borough President Connor informed the electric company that the islands were designated to become parks. The chairman of Con Edison stated, "Nuclear stations have a perfect safety record thus far."[293] Apparently the idea was still on the table in 1968, when Con Ed announced possible plans for a nuclear generating station on a man-made island off Staten Island's South Shore if the need for additional electricity still existed seven to twenty years in the future. An atomic generator would also provide fresh drinking water by powering a desalination plant. To create the island, Con Ed suggested compacting garbage—a "commodity" that was readily available on the island at the time.[294]

In 1968, a modern policy of placing electrical wires underground when possible was revealed by the utility company. This would occur in developments of five or more "reasonably contiguous" homes. Installing electrical lines underground was ten to fifteen times more expensive than mounting overhead wires, but the company said it hoped this improvement would benefit neighborhoods.

Modern Con Ed also announced that coal with a lower sulfur content would soon be burned.[295] A proposal for all-electric homes was also part of Con Edison's promotions in 1968. The plan was called "total electric living." Customers would receive reduced rates for electric heating and hot water during the winter, while summer rates would increase. Privately owned homes and multiple dwellings would benefit. The plan would not be available for apartment buildings, schools, or office buildings. It was approved by the Public Service Commission.[296]

Brooklyn Union Gas foresaw continued success in the field of customer relations in 1969. One concern was that gas would not be connected to as many Staten Island houses as in the past. It seems that construction was down since a septic tank ban had gone into effect on the island. It was projected that 1,200 fewer homes would be built on the island owing to the prohibition.

Brooklyn Union Gas began Staten Island service in 1957. By 1960, natural gas was mainly used for heating purposes in private homes and larger buildings. More than one thousand homes and businesses had converted to gas on Staten Island in 1959. Brooklyn Union predicted that more than two-thirds of existing homes in 1960 would be heated by natural gas by 1970.[297]

To promote and encourage use of natural gas, Brooklyn Union touted its Blue Star model home on Old Town Road in Grasmere in 1964. A cooperative effort between the company and the Staten Island Home Builders Association, this one-family high ranch was equipped with a furnace, air-conditioning, a water heater, a refrigerator, cooking facilities, a clothes dryer, and outdoor illumination that all functioned on gas. The model was open to the public for viewing.

Chapter 23

OVERDEVELOPMENT

While many residents believe Staten Island is overdeveloped, the City of New York has continually encouraged the borough's population to increase. It's simple: More residents and businesses bring more money into the city's coffers. Even so, there have been ample discussions about who is to blame for the overdevelopment and lack of planning. Many faulted the people from Brooklyn who resettled on Staten Island. Others blamed the politicians, still others the developers. For the most part, laws to protect the environment did not exist or were not enforced before the 1970s. Laws to prevent overpopulation did not exist. Laws to prevent changing the topography did not exist. The natives of Staten Island were soon complaining about the changes taking place. Soon, many of the folks from Brooklyn were grumbling. No one was really listening to either group.

The election of the next borough president of Staten Island took place in early November 1964, the same month that the Verrazzano-Narrows Bridge opened. The main contenders were Albert Maniscalco and Robert Connor. Maniscalco had served in the position for the previous fifteen years. It was a time of tumult and change like no other in the history of Staten Island. Maniscalco lost. The day after the election, November 4, 1965, it was said he was more relaxed than anyone had ever seen him. By his own admission, he slept better the previous night than he had in a long time. As the night wore on, he was aware that Connor had won the election. The outgoing borough president arrived at Borough Hall at his customary time of seven-thirty in the morning. He spent most of the morning cheering up aides and employees who were "stunned" that he had suffered defeat. Many of them were in tears. Maniscalco had not considered future plans as he believed that he would win reelection. But it seems that the borough wanted a change.[298]

When Albert Maniscalco died on September 2, 1998, the *New York Times* noted that he was borough president of Staten Island from 1954 to 1965, a time that saw a "dizzying shift from [a] rural to [an] urban borough." His obituary cited a 1965 quote from Ada Louise Huxtable, who wrote, "Development was racing forward on Staten Island with a haphazard abandon that could not be better calculated to destroy the countryside if it had been planned by enemy action." Huxtable later applauded Maniscalco for "delaying an urban renewal project that would have accelerated the damage."[299] It was also recalled that Maniscalco stood firmly opposed to the original route of the Richmond Parkway since it would have ripped through the Staten Island Greenbelt—a term he was credited with developing when he first used it at a Board of Estimate meeting.

By 1970, certain sections of Staten Island were no more familiar to natives than the landscape of Saturn. One- and oftentimes two-family homes on crowded streets that lacked appropriate parking were the norm. As the new decade arrived, residents noticed that—in modern neighborhoods, at least—telephone poles had vanished as phone lines were now underground. Here and there, one could also see private sewage plants for certain developments since the island's main interceptor sewer was far behind schedule. Economics was also changing the preferred type of structure now put up by builders. Two-family houses were selling well as the availability of mortgages had improved. Obviously, the rental income was vital as the cost of houses increased. The *New York Times* reported in 1971, "Separate houses are continuing to rise in rigid formation across the current major development area—the midsection of the island." Resistance to cluster residences continued as the majority of housing lots were the usual forty by one hundred feet. Each had its "tiny new tree" with a "perfect new lawn of greener-than-green sod, with no sign of a weed. And each has a recent buyer." The reporter contended that the robust demand for houses led to a lack of "variation in architectural design and land planning." Indeed, a builder in Eltingville was building seventy two-family houses off Ridgewood Avenue. Sixty-five of the houses were already sold even though only thirty-five were under construction. None were complete.[300]

During the 1960s, rampant development led to the destruction of many historic structures. Today, 1922 Arthur Kill Road is an empty site near the planned community known as Village Greens. The development is bounded by Woodrow Road, Arden Avenue, and Huguenot Avenue. In 1969, the Reverend Orin Griesmyer and his wife lived in a pre-Revolutionary home that was attached to a Second Empire house. The building was known as the Cropsey House (as in the early Staten Island family, not the killer of urban legend). Joost Cropsey came to America from Germany via Holland in 1652, settling first in Long Island. Some of his descendants moved to Staten Island. One descendant was the famous Hudson River School painter Jasper Cropsey. By the late 1920s, Laura Cropsey was living in the house.

The Griesmyers adored their home and the acre it sat on. They were thrilled with the wildlife they encountered. Unfortunately, the couple would soon be surrounded by 1,200 homes. The Reverend Griesmyer was pleased that streets would follow land contours and that watercourses and wetlands would not be diverted. Griesmyer knew that change was unavoidable. He was hopeful that New York City red tape would bog down the commencement of construction.[301] It did not.

IN CONCLUSION

According to the New York City Department of Sanitation, the population of Staten Island passed 300,000 during early March 1967. The exact number was 301,052. The department arrived at this figure by multiplying the number of newly occupied dwelling units by the national family average of 3.9 persons. According to experts, the Sanitation numbers were "amazingly accurate." The department determined that by 1969, Staten Island's population would be 350,000.[302]

By 1969, 80 percent of Staten Island families owned the home where they resided. The non-white population was between 7 and 10 percent.[303] While the majority of citizens were registered Democrats, they generally voted for Republican candidates. In 1969, Staten Island had ninety schools that served approximately fifty thousand children, 40 percent of whom went to parochial schools.

Finally, in 1970, a master plan for Staten Island was revealed. Released by the New York City Planning Commission on March 2, it was called a "Plan for New York." It was primarily a "strategy for coherent development with heavy emphasis on the relationship between transportation and development." It was not a plan for "laying out communities, commercial centers and a street network in great detail."[304] How could it have been? After all, many of the communities, commercial centers, and street networks were already laid out. The "plan" was sixty-six pages of information on zoning, mapping, financial decisions, and site selections, filled from front to back with facts, figures, maps, statistics, and charts.

Many recommendations were to be found in the document, including a suggestion that the North Shore Staten Island Rapid Transit system be resurrected. Another proposal, submitted the previous decade, was to landfill the waters between Swinburne Island, Hoffman Island, and South Beach. It would be accomplished by using that fascinating oxymoron "clean fill"—in other words, garbage. The clean fill would be dumped along the shoreline. It would double the land area for residential and recreational use. The plan envisioned communities of 250,000 to 500,000 people for District 4, which held 6,500 acres of vacant land. This location was the previously mentioned South Richmond. The district would offer employment opportunities, recreational facilities, and modern transportation. The needs of residents would be filled.[305]

There was the usual call for cluster housing with accompanying open space and varied house patterns. Perhaps the most surprising suggestion—and one that was well-thought-out owing to the side effects of the rampant unplanned development of the 1960s—was a call for the preservation of existing well-maintained housing. Even so, demolition of original housing would soon become a common occurrence on the island as land became scarce for new construction in the 1980s. These planners realized that much of the pre-bridge housing was located on oversized lots, while old estates and institutions with vast acreage still existed on the island. When all the vacant space was built on, the developers would soon buy sizable lots for sizable prices. The original home would then be destroyed, no matter if it was a stunning nineteenth-century Victorian with its original architectural décor intact or a pre–Civil War row house with historic lineage. Rows of townhouses and condominiums would soon spring up like cornstalks across the borough. In most cases, they were accompanied by inadequate driveways and scarce street parking (in order to cram as many structures as possible onto a plot that originally held a one-family home). The new homeowner might have a sewer, but the age-old Staten Island problem of flooding was frequent and rampant. The makeup of older neighborhoods like Mariner's Harbor, Travis, Tottenville, Rossville, and others would be forever altered.

Thanks to the vigilance of the Westerleigh Improvement Society, that community was spared much of the exploitation that was heaped on the previously mentioned neighborhoods. In a few cases, rezoning helped preserve some of the older housing stock, but rezoning was always a poor tool for preservation. The only safeguard for older homes in established neighborhoods was New York City landmarking, which the majority of

Staten Islanders were and still are against. They believed a landmarked home was a home that could not be altered or resold without Big Brother's authorization. Many one-family houses ended up in the Fresh Kills Landfill only to be replaced by multiple housing units and townhouses that "caterpillared" around narrow, car-lined streets.

In 1972, you could still buy manure on Staten Island. A sign on Richmond Avenue in New Springville read "Manure. All you want. $5." There were three riding stables, four colleges—two were actually free—a gypsum plant, a soap factory, and a living history village, but things had changed. Did they go from bad to worse? That is up to the individual.

In 1972, Weissglass Speedway had just closed. Geraldo Rivera was about to report nationally on the horrendous conditions at Willowbrook State School after an *Advance* exposé. The previous year, riots took place at Port Richmond High School. In a few years, the historic community of Sandy Ground would be spliced by the West Shore Expressway, while Sailors Snug Harbor would soon be vacated after almost 140 years of operation. The Snugs were being transported to the less expensive landscape of North Carolina.

Sailors' Snug Harbor in Livingston was home to thousands of former sailors, merchant marines, and ship's captains. *Undated photo, Library of Congress.*

In the mid-1970s, Staten Island had families, many of whom were newcomers or were made to feel like newcomers. Some went to South Beach, but the water was still full of oil. There was an amusement park called Kiddie Town that was enjoyed by all. In a bar on Main Street in Tottenville known as Club 93, such tunes as "A Cowboy's Work Is Never Done" wailed from a jukebox by the wall. Main Street was also the location for Tottenville Horse Service and Supply. A journalist standing at Main Street and Amboy Road wrote, "For nearly five minutes nothing moves there." This astute documentarian also noted that on Staten Island, "poor people live at the bottom of hills, and money goes upward in ascending layers."[306] On Staten Island, rapid transit trains' seats were still made of wicker that the conductor could flip back and forth according to the direction of train travel. Of course, in the mid-1970s, the conductors still wore shiny silver badges as they went from seat to seat to collect passengers' fares as each boarded the train.

Train service and Tottenville might not have changed much by the mid-1970s, but the rest of Staten Island had. It would not be long before development arrived at the far reaches of the borough. Tottenville, Rossville, Sandy Ground, Greenridge, and Charleston were about to be crushed under development. Sections of the South and West shores had been ignored by many developers during the rush of building after the opening of the bridge. This had much to do with the lack of sewers and Department of Health rules on the installation of septic tanks. But the opening of the West Shore Expressway and the construction of private sewer facilities in developments would drastically alter these once-countrified communities.

Bethlehem Steel shut its Staten Island shipyard in the 1960s. Only a propeller plant remained. It closed in 1972. Factories like S.S. White and U.S. Gypsum trudged into the 1970s, but by 1980, only Procter & Gamble, Nassau Recycling Corporation, and Superior Chocolate were still in business. U.S. Gypsum had been the target of neighborhood complaints for years. In the fall of 1965, a meeting was facilitated by plant managers and neighbors to discuss the issues. Even the district attorney was present. According to one nearby resident, pollution from Gypsum's stacks was so severe that his children couldn't sleep for coughing. A Gypsum rep stated the company would review its procedures but, as far as he knew, the plant conformed to all regulations. Even so, the company wanted to be a good neighbor. The New Brighton concern employed six hundred individuals and had a weekly payroll of $100,000.[307] The setup officially closed in 1976. The reasons given were lack of space to expand, an aging plant, and increased traffic on Richmond Terrace.

Superior Chocolate, now known as Supreme Chocolate, is today the last manufacturer of any major product on Staten Island. In other words, one criterion for the opening of the Verrazzano-Narrows Bridge was not met: The bridge did not spur industrial growth.

The next phase of development on Staten Island saw developers buying one-family houses built in the developments of the late 1950s and early 1960s. In many cases, they also bought the lots on either side of the house. All three houses might be demolished, or perhaps a remnant of one would be left standing. Then an enormous and oftentimes outlandish structure would be built to replace the original home. Oftentimes, these larger buildings, especially if they were ornate, were referred to as McMansions.

In 1990, the population of Staten Island was said to be 378,977. Local officials claimed this was an undercount. With the turning of the millennium, Staten Island's population had risen to 443,728.[308] In July 2024, the population of Staten Island was estimated to be 498,212.[309]

There are still septic tanks, but not many. There is still flooding. There are still overcrowded schools. The Gericke Farm in Rossville exists. It had been purchased for inclusion into Clay Pit Ponds State Park Preserve by 1980. The Sylvanus Decker farm owned by Historic Richmond Town, lovingly worked by Tim Anagnostis until a few months before his death on August 31, 2000, exists also. Many acres of Staten Island land were converted to parkland thanks to the early efforts of certain foresighted individuals and organizations, including William T. Davis and what is now the Staten Island Museum. In order to preserve Clay Pit Ponds State Park, a powerhouse organization was formed. Protectors of Pine Oak Woods would go on to preserve Blue Heron Park, sections of the Greenbelt, Bluebelt areas, Kingfisher Park, and acres of additional open space. Over the years, Protectors has been guided by several stalwart officers and members. These include but are not limited to Richard Buegler, Ellen Pratt, Ed Milanese, Carlton Beil, Herman and Syl Zaage, Elaine Crouteau, Cliff Hagen, Eileen Monreale, and many others too numerous to mention. The group continues to forge on.

The Richmond Parkway still exists on official New York City maps. Protectors of Pine Oak Woods has been stressing this fact for years. Its presence on these maps makes it a threat to the Greenbelt. It needs to be removed. People need to be vigilant. An event in 2012 cheered naturalists. An exit from the Staten Island Expressway that stood next to the Michael Petrides School for decades was demolished and removed. It was meant to lead drivers through the Greenbelt along the Richmond Parkway. The roadway would have cut through Todt Hill and Pouch Camp and clipped portions of High Rock and La Tourette Parks.[310] Mayoral candidate John Lindsay suggested to Robert Moses that he change the route of the parkway to prevent environmental destruction. Moses refused. In fact, some of the serpentine rock removed for construction of this exit was carted away and dumped on the northwest corner of Manor Road and Rockland Avenue in the Greenbelt. This sky-high deposit was to have been used for a connection between the Willowbrook and Richmond Parkways. It was spitefully named Moses Mountain for its upsetting presence in this pristine location. In recent years, it was officially renamed Mount Paulo to honor the late Staten Island Parks Commissioner Thomas Paulo, who did so much to preserve and tend to the environmental beauty of the borough.

With the opening of the Verrazzano-Narrows Bridge in 1964, Staten Island lost many things. Chief among them was isolation from the rest of New York City. This sparkling modern pathway had opened the doorway to the island, and that doorway would never close again. The sense of "islandness" experienced by pre-bridge dwellers was amazing for how long it lasted. What is missed? Certainly not the dirt roads, the mass of septic tanks, and the lumbering ferry that was needed by so many to reach Manhattan. What did happen—and it continues to happen—is the continuous blame placed on "newcomers." No one can stop change or the arrival of modernity. Whether it is immigrants, technology, or a change in the governing political party, someone or something must be pointed to as the cause for that change. Change is feared, sometimes rightly so. We are upended by the unfamiliar, the things we do not know or understand. We understood the open treed lot that stood for decades. Many do not understand or take a moment to understand the arrival of something different. Many have also forgotten the need to speak up when something is fundamentally wrong. We can learn much from Protectors of Pine Oak Woods, the National Association for the Advancement of Colored People, the Westerleigh Improvement Society, and others.

NOTES

Chapter 1

1. Philip J. Brittain, "Stores, Homes Take Over Where Farms Flourished," *Staten Island Advance*, October 8, 1964.

Chapter 3

2. "Borough of Richmond" ad, *Sun*, July 1, 1906.
3. Advertisement in the *New York Tribune*, June 26, 1921.
4. Advertisement in the *New York Tribune*, June 26, 1921.
5. "Road System Booms Staten Island Sales," *New York Times*, April 8, 1928.
6. "$60,000,000 Bridge Planned Over Narrows," *New York Herald Tribune*, May 2, 1928.
7. "Span of 4,500 Feet in Narrows Bridge," *New York Times*, November 17, 1929.
8. Robert Moses, letter to the editor, *New York Times*, April 19, 1956.

Chapter 4

9. James Whitford, "Property Auction Bring Barbs Along with Cash," *Staten Island Advance*, April 18, 1965.
10. "City to Sell Island Land on Thursday," *Staten Island Advance*, June 13, 1965.

11. Peter Kihss, "Favoritism Laid to Borough Head of Staten Island," *New York Times*, January 26, 1960.
12. Phillip Dougherty, "Staten Island Pushed into a New Era," *New York Times*, November 21, 1964.
13. Staten Island Home Builder's Association, "Staten Island Home Buyers Guide," circa 1962. Collection of the Staten Island Historical Society.

Chapter 5

14. "Bridge to Become All-Moses Operation," *Staten Island Advance*, December 12, 1959.
15. S.S. McSheehy, "3,000 Watch History in the Making," *Staten Island Advance*, August 14, 1959.
16. Robert Caro, *The Power Broker* (Vintage Books, 1975), 736.
17. Mark Weisner, "Narrows Bridge Claims First Victim as Worker Plunges to Death," *Staten Island Advance*, August 25, 1962.
18. Wikipedia, "Verrazzano-Narrows Bridge," https://www.wikipedia.org; Gay Talese, *The Bridge: The Building of the Verrazano-Narrows Bridge* (Bloomsbury, 2014), 106.
19. "300-Foot Fall Off Bridge Kills Youth," *Staten Island Advance*, October 10, 1963.
20. "Bridge Crews Resume Monday," *Staten Island Advance*, December 7, 1963.
21. James S. McLemore and James P. McLemore, "Bridge-Building Family: 'It's Windy Up There,'" *Staten Island Advance*, April 27, 1964.
22. Structurae, "Othmar Ammann," http://structurae.net.
23. Caro, *Power Broker*, 843.
24. Murray Schumach, "Verrazano Bridge Casts Long Shadow over Staten Island," *New York Times*, June 28, 1969.
25. Robert J. Popp, "Bridge Builders Make Lab of Tiny Island," *Staten Island Advance*, March 13, 1959.
26. Alfred G. Haggerty, "Former Home of Generals Coming Down," *Staten Island Advance*, August 18, 1959.
27. Nadia H. Youseff, *Population Dynamics on Staten Island* (Center for Migration Studies, 1991), 58.
28. S.S. McSheehy, "Bridge Boosters Sound Call for Island Help at Hearing," *Staten Island Advance*, June 5, 1958.
29. S.S. McSheehy, "Bridge Road Talkathon Opens Way for Ruling," *Staten Island Advance*, July 1, 1958.

30. S.S. McSheehy and Stanley Terkelsen, "City Hall Holds Up OK on Bridge," *Staten Island Advance*, September 27, 1958.
31. Robert Moses, "Staten Island Can Be Promised Land," *Ogdensburg Advance News*, September 5, 1965.
32. Moses, "Promised Land."
33. Moses, "Promised Land."

Chapter 6

34. Ellen Fletcher, "Verrazano-Narrows Bridge," in *Encyclopedia of the City of New York*, edited by Kenneth L. Jackson (Yale University Press, 2010), 1366.
35. "View on TV: Paradise for Speculators," *Staten Island Advance*, November 18, 1964.
36. "TV View of Our Island an 'Evil Eye'—Moses," *Staten Island Advance*, November 21, 1964.
37. "TV View."
38. "Ceremonies Tomorrow Open an Era," *Staten Island Advance*, November 20, 1964.

Chapter 7

39. "Company Faced Rocky Road," *Staten Island Advance*, November 20, 1964.
40. "They Paved the Way Across," *Staten Island Advance*, November 20, 1964.
41. "They Paved the Way."
42. "$6,978 Bid to Raze 12 Buildings," *Staten Island Advance*, September 16, 1964.
43. "S.I. Group Seeking Shift of Parkway," *New York Times*, January 16, 1966.
44. Cy Berlowitz, "Park Boss Leads Trek in White Green Belt," *Staten Island Advance*, January 30, 1966.
45. Robert Moses, "Our Future's Bright, Moses Report Says," *Staten Island Advance*, August 15, 1965.
46. "Car Traffic Doubles at Goethals Bridge," July 3, 1965.
47. Fletcher, "Verrazano-Narrows Bridge."
48. Diane C. Lore, "From 50 Cents to $16," *Staten Island Advance*, November 17, 2014.
49. Tom Wrobleski, "Skimpy Attendance, Standard Complaints Turns MTA Hearing into Ground Hog's Day," *Staten Island Advance*, December 12, 2014.

50. Metropolitan Transit Authority, "New Toll Rates at MTA Bridges and Tunnels—Implementation August 6th, 2023," updated August 7, 2024, https://www.mta.info.
51. Leonard Novarro, "Parking Facilities Key to Stapleton's Problem," *Staten Island Advance*, June 21, 1965.

Chapter 8

52. This chapter's title is from Ada Louise Huxtable, "Staten Island's Beauty Losing to Builders," *New York Times*, August 9, 1965.
53. Bernard Stengren, "Bridge Brings Boom to Bucolic Staten Island," *New York Times*, October 2, 1963.
54. James Whitford, "Builder Has a Voice, but the Buyer Is Seldom Heard," *Staten Island Advance*, August 12, 1965.
55. "'65 Total Passes 850," *Staten Island Advance*, April 3, 1965.
56. Saving.org, https://www.saving.org.
57. Stengren, "Bridge Brings Boom."
58. Denis LeBlanc, "Building Now Largest Industry," *Staten Island Advance*, April 25, 1964.
59. LeBlanc, "Building Now Largest Industry."
60. LeBlanc, "Building Now Largest Industry."
61. Denis LeBlanc, "Staten Island Builders Wide Variety in Styles, Price," *Staten Island Advance*, May 15, 1965.
62. LeBlanc, "Staten Island Builders."
63. LeBlanc, "Staten Island Builders."
64. "Deputy Mayor Pays Heed to Complaints by Island Builders," *Staten Island Advance*, March 4, 1965.
65. "Deputy Mayor Pays Heed."
66. "Prognosis Good," *Staten Island Advance*, March 6, 1964.
67. Harold Birns, "Growth to Explosive Proportions Is Inevitable," *Staten Island Advance*, April 24, 1965.
68. James Whitford and Drew Fetherston, "'Man on the Street' Looks at His Growing Home Town," *Staten Island Advance*, October 5, 1965.
69. Robert Moses, "Remarks of Robert Moses at the Celebration of the Seventy-Fifth Anniversary of the Staten Island Institute of Arts and Sciences," April 13, 1957. From the History Archives and Library of the Staten Island Museum.
70. Steven V. Roberts, "The Development of Staten Island: Will It Become Another Queens," *New York Times*, April 23, 1967.

71. Lloyd B. Tamarin, "Staten Island Building Defended," letter to the editor, *New York Times*, August 23, 1965.
72. Drew Fetherston and James Whitford, "Builders: 'Want Planning Voice,'" *Staten Island Advance*, October 4, 1965.
73. Fetherston and Whitford, "Builders: 'Want Planning Voice.'"
74. Fetherston and Whitford, "Builders: 'Want Planning Voice.'"
75. Drew Fetherston and James Whitford, "Profit-Hungry Builders Ruining Island, Critics Cry," *Staten Island Advance*, October 3, 1965.
76. "Farewell to SI?" February 17, 1966.
77. "Staten Island Down the Drain," *New York Times*, January 22, 1966.
78. Ada Louise Huxtable, "Staten Island's Beauty Losing to Builders," *New York Times*, August 9, 1965.
79. Huxtable, "Staten Island's Beauty."
80. Huxtable, "Staten Island's Beauty."
81. "Ruining Staten Island?" *New York Times*, August 10, 1965.
82. Mark Weisner, "City Planner Blasts Hill Developers," *Staten Island Advance*, October 4, 1966.
83. Weisner, "City Planner Blasts."
84. "Citizens Group Foresees Plan Headaches on Island," *Staten Island Advance*, December 19, 1964.
85. Drew Fetherston, "Dumping Can Be Problem, Especially for Contractors," *Staten Island Advance*, May 2, 1965.
86. James Whitford, "Master Plan a 'Crucial' Need," *Staten Island Advance*, February 1, 1965.
87. Drew Fetherston and James Whitford, "All Agree on Planning Need, but Views Differ on Methods," *Staten Island Advance*, October 8, 1965.
88. Fetherston and Whitford, "All Agree."
89. Mark Weisner, "Urge Island Development Boss, South Shore Control Agency," *Staten Island Advance*, June 12, 1966.
90. Weisner, "Urge Island Development Boss."
91. "Meyer's Swearing-In Set," *Staten Island Advance*, August 3, 1966.
92. "Planner Is Chosen for Staten Island," *New York Times*, August 4, 1966.
93. S.S. McSheehy, "The Borough President: 'Mayor Faces a Fight,'" *Staten Island Advance*, August 5, 1966.
94. McSheehy, "Borough President."
95. "Lindsay Heads for Defeat on Meyer Office," *Staten Island Advance*, October 27, 1966.
96. "Meyer Tells How He'll Use Funds," *Staten Island Advance*, April 16, 1967.
97. "Meyer Picks 14-Member Advisory Unit," *Staten Island Advance*, January 25, 1967.

98. Drew Fetherston, "Zoning Concept Good, But Faulty," *Staten Island Advance*, February 1, 1967.
99. Fetherston, "Zoning Concept Good."

Chapter 9

100. Denis LeBlanc, "Builders Feeling Impact," *Staten Island Advance*, April 24, 1965.
101. Harold Birns, "Island Sets Pace in Building Race," *Staten Island Advance*, April 25, 1964.
102. "1.5 Million Homes Predicted," *Staten Island Advance*, October 23, 1965; "Island Developers Boost '65 Building Plans Total," *Staten Island Advance*, January 8, 1966.
103. Leonard Novarro, "Slump in New Building Exacting a Heavy Price," *Staten Island Advance*, October 9, 1966.
104. "Choice Land on Island Disappears," *Staten Island Advance*, April 1, 1967.
105. "Foresee Comeback in 1968," *Staten Island Advance*, February 4, 1968.
106. "Foresee Comeback in 1968."
107. Roberts, "Development of Staten Island."
108. Roberts, "Development of Staten Island."
109. Roberts, "Development of Staten Island."
110. James Whitford, "Better Times Ahead for Island Builders," *Staten Island Advance*, April 30, 1967.
111. Richard Nusser, "Mayor Hopes Island Could Be Ideal Suburb," *Staten Island Advance*, February 2, 1969.

Chapter 10

112. Denis LeBlanc, "Builders Extend Warm Welcome," *Staten Island Advance*, September 19, 1964.
113. Albert V. Maniscalco, "Borough President Views Island Role in Progress Pageant," *Staten Island Advance*, April 25, 1964.
114. "Councilman Asks Study of Island Transit," *Staten Island Advance*, June 23, 1965.
115. Drew Fetherston and James Whitford, "Nation Eyes Island as Test Case for Planners," *Staten Island Advance*, September 29, 1965.
116. Roberts, "Development of Staten Island."

117. "C of C Urges City to Map Streets," *Staten Island Advance*, March 1, 1966.
118. Mark Weisner, "Unlawful Homes Reach 1,800," *Staten Island Advance*, November 13, 1966.
119. "'Life in 1975, Will Be Quite Different,' Tobin Warns," *Staten Island Advance*, February 15, 1955.
120. Drew Fetherston and James Whitford, "Checkerboard Street Pattern—Legacy of a Boom," *Staten Island Advance*, September 30, 1965.
121. Ada Louise Huxtable, "Staten Island's Beauty Losing to Builders," *New York Times*, August 9, 1965.
122. Huxtable, "Staten Island's Beauty."
123. "Staten Island Gets New Apartments," *New York Times*, April 14, 1963.
124. "Mason Ave. Families Picket Street-Widening Project," *Staten Island Advance*, August 13, 1965.
125. Carol Berman, "Housewives and Babies Protest," *Staten Island Advance*, August 8, 1967.
126. Berman, "Housewives and Babies Protest."
127. "Mothers, Children Picket Meyer Home," *Staten Island Advance*, September 7, 1967.
128. "Picket My Office and Picket They Did," *Staten Island Advance*, September 13, 1967.
129. Louis Sperksky, "Legal Grading Can Turn Roadway into a Wall," *Staten Island Advance*, December 19, 1965.
130. William Robbins, "Staten Island Neighbors Plead with City Not to Bury Home," *New York Times*, February 17, 1966.
131. Robert Foster, "Residents Prepare Battle Against Sunken Homes," *Staten Island Advance*, October 7, 1965.
132. Joseph Margolis, interview with Lois Gilman, July 15, 1975. Accessible in the Archives of the Staten Island Institute of Arts and Sciences, Box 83.88T.
133. "Health Hazard Is Feared in Oakwood Flood Spots," *Staten Island Advance*, June 25, 1967.
134. Jerry Campo, "D.A. Starts Probe of New Dorp Flood," *Staten Island Advance*, August 8, 1967.
135. Jerry Campo, "Emergency Steps Taken on Tysens Lane," *Staten Island Advance*, August 29, 1967.
136. "Crane to Start Dredging in Oakwood Today," *Staten Island Advance*, September 15, 1967.
137. "Crane to Start Dredging."
138. S.S. McSheehy, "Helpless About Dumping on Streets, City Admits," *Staten Island Advance*, February 12, 1964.

139. S.S. McSheehy, "Planning Agency Eases Street-Width Restrictions," *Staten Island Advance*, October 8, 1966.
140. "Below-Grade Streets Receive Mayor's OK," *Staten Island Advance*, October 10, 1967.
141. William Tow and George O. Reddy, "41 Rescued in Beach Floods," *Staten Island Advance*, January 24, 1966.
142. Tow and Reddy, "41 Rescued."
143. Drew Fetherston, "Flood Victim Got Answer from City," *Staten Island Advance*, August 9, 1967.
144. Drew Fetherston, "City Checking Sewer Link-up," *Staten Island Advance*, August 10, 1967.
145. Raymond Wittek, "No Action by Jury on Flood," *Staten Island Advance*, December 21, 1967.
146. Wittek, "No Action by Jury."
147. "Men Form Rain Patrol in Oakwood," *Staten Island Advance*, August 29, 1967.
148. James Whitford, "Abandoned Site Alarms Residents," *Staten Island Advance*, June 21, 1967.
149. George O. Reddy, "2 Young Girls Drown in Pond," *Staten Island Advance*, January 11, 1966.
150. Drew Fetherston, "Pond Probe Pledged," *Staten Island Advance*, January 12, 1966.
151. "Full-Fledged Jury Probe Vowed in Girl's Death," *Staten Island Advance*, January 13, 1966.
152. "Full-Fledged Jury Probe."
153. Raymond Wittek, "Bare Pond Trap Violations," *Staten Island Advance*, January 18, 1966.
154. Wittek, "Bare Pond Trap Violations."
155. "Settlement Brings an End to Suits in Two Drownings," *Staten Island Advance*, May 20, 1967.
156. "Settlement Brings an End."
157. "Pond Deaths Testimony Satisfactory, Judge Rules," *Staten Island Advance*, November 17, 1966.
158. "Realty Men Enter Innocent Pleas in Girls' Drownings," *Staten Island Advance*, April 16, 1966.
159. "Court Frees Defendant in Drownings," *Staten Island Advance*, January 11, 1967.
160. "Settlement Brings an End."

Chapter 11

161. "Builders Lose Septic Test Court Fight," *Staten Island Advance*, September 2, 1965.
162. Steven V. Roberts, "City Stiffens Septic Tank Regulations," *New York Times*, December 3, 1967.
163. Richard Amberg, "Sewer Funds on Paper Seldom Turn to Cold Cash," *Staten Island Advance*, September 16, 1968.
164. Richard Amberg, "One Sewer Completed on North Shore in 15 Years," *Staten Island Advance*, September 23, 1968.
165. "Septic Tank Ban," *Staten Island Advance*, January 18, 1970.
166. David O. Stewert, "Septic Tanks: A Constant Drain," *Staten Island Advance*, June 13, 1972.
167. James Whitford, "Planners Push Heartland Sewer," *Staten Island Advance*, May 12, 1966; "Planners OK Sewage Plant for Great Kills," *Staten Island Advance*, December 1, 1966; S.S. McSheehy, "Private Sewer System Called Only Answer," *Staten Island Advance*, June 7, 1966.

Chapter 12

168. "Total Population. New York City and Boroughs, 1900–2010," https://www.nyc.gov/assets/planning.
169. Nadia H. Youseff, *Population Dynamics on Staten Island* (Center for Migration Studies, 1991), 1.
170. Michael Azzara, "CORE Chief Says 'Neutrals' Are Segregation's Allies," *Staten Island Advance*, June 3, 1964.
171. Azzara, "CORE Chief Says."
172. Claire Regan, "Richard Prideaux, a 1960s Force for Racial Equality on Staten Island, Dies. He Was 94," *Staten Island Advance*, March 7, 2021.
173. "CORE Plans Outlined at Meeting," *Staten Island Advance*, July 14, 1961.
174. "Anti-Bias Pickets Head for 2 Island Schools," *Staten Island Advance*, July 22, 1963.
175. "NAACP to Join CORE for Bigger Push on Island," *Staten Island Advance*, July 23, 1963.
176. Mark Wiesner, "Bias Blockade at School Jails 5," *Staten Island Advance*, July 25, 1963.
177. "Bias Pickets Returns Seek Sit-down Arrests," *Staten Island Advance*, July 24, 1963.

178. "Bias Protest Hits Borough Hall," *Staten Island Advance*, July 26, 1963.
179. "Will Help on Rights, B.P. Tells Delegation," *Staten Island Advance*, July 27, 1963.
180. Regan, "Richard Prideaux."
181. "Realty Bias Charged, but Firm Denies It," *Staten Island Advance*, October 12, 1963.

Chapter 13

182. "New 'Sinking' Rumors Engulf Huguenot JHS," *Staten Island Advance*, February 11, 1965.
183. Kenneth R. Fanizzi, "High School at Miller Field Takes a Giant Step," *Staten Island Advance*, February 11, 1965.
184. Erwin Endress, Staten Island Advance, "First Portables Becoming Reality at PS 42," *Staten Island Advance*, July 18, 1963.
185. "Portables Promised in Time for Opening Day of School," *Staten Island Advance*, August 7, 1964.
186. "PS 52 to Open 7 Months Ahead of Schedule," *Staten Island Advance*, November 13, 1964.
187. Kenneth Fanizzi, "B.P. Speeds Portables as Pickets March," *Staten Island Advance*, September 1, 1964.
188. Jack Reycraft and Philip Brittain, "PS 8 Boycott Hits 50%," *Staten Island Advance*, September 14, 1964.
189. "PS 8 Boycott Holds Tight," *Staten Island Advance*, September 15, 1964.
190. "2 Island High Schools Put in Integration Plan," *Staten Island Advance*, January 25, 1964.
191. Kenneth R. Fanizzi, "Housewives' Census Backs Demand for New Schools," *Staten Island Advance*, February 26, 1965.
192. Kenneth R. Fanizzi, "Crowded Schools? Officials' Survey Finds Empty Seats," *Staten Island Advance*, March 28, 1965.
193. "Modernization Bureau Set Up," *Staten Island Advance*, February 9, 1965.
194. Kenneth R. Fanizzi, "RX for Old Schools: $13 Million Fixup," *Staten Island Advance*, June 10, 1966.
195. Kenneth R. Fanizzi, "Little PS 15 Just Wants to Be Let Alone," *Staten Island Advance*, February 23, 1965.
196. Kenneth R. Fanizzi, "PS 15 Kids' Parents Gird for War on Shift to New School," *Staten Island Advance*, March 17, 1965.

197. "Early Demolition Due PS 3, City's Last Wooden School," *Staten Island Advance*, June 28, 1965; McCandlish Phillips, "It's Vacation Time in Public Schools, *New York Times*, July 1, 1965.
198. "Public Hearing Slated on School Projects," *Staten Island Advance*, May 11, 1964.
199. Kenneth R. Fanizzi, "Concord Moms to March in Protest of School Lag," *Staten Island Advance*, August 15, 1965.
200. Fanizzi, "Concord Moms to March."
201. James B. Donovan, "Board of Education Striving for 'Quality Growth,'" *Staten Island Advance*, April 24, 1965; Fanizzi, "Concord Moms to March."
202. Kenneth R. Fanizzi, "300 Island Parents Voice Plea for Better Schools," *Staten Island Advance*, February 18, 1966.
203. Kenneth R. Fanizzi, "Island Public Schools May Enroll Record 40,644," *Staten Island Advance*, September 6, 1966.
204. Kenneth R. Fanizzi, "64,265 Heading Back to Classrooms," *Staten Island Advance*, September 11, 1966.
205. Kenneth R. Fanizzi, "Schools Short on Teachers," *Staten Island Advance*, September 9, 1966.
206. Kenneth R. Fanizzi, "Seek Local Rule for Pupil Busing," *Staten Island Advance*, October 31, 1969.
207. Kenneth R. Fanizzi, "St. George, Civic and Transport Hub, Also Center of Student Information," *Staten Island Advance*, October 8, 1964.
208. Kenneth R. Fanizzi, "It's Cars vs. Kids in St. George," *Staten Island Advance*, December 19, 1966.
209. "Public School Enrollment Shows Gain," *Staten Island Advance*, January 21, 1968.
210. Kenneth R. Fanizzi, "Proposed PS 54 May Seat 1,200," *Staten Island Advance*, March 16, 1967.
211. "Portables Substitute for New School," *Staten Island Advance*, February 27, 1969.
212. Kenneth R. Fanizzi, "Islanders Press for 6 New Schools," *Staten Island Advance*, September 24,1969.
213. Fanizzi, "Islanders Press."
214. Kenneth R. Fanizzi, "School Board Honors Negro in IS 61 Name," October 28, 1969.
215. Erwin Endress, "Assumption School Opens to First 100 Pupils," *Staten Island Advance*, March 20, 1962.
216. "Catholic Schools Enroll 22,218," *Staten Island Advance*, September 17, 1965.

217. "Bingo Spells Profit for Island Groups," *Staten Island Advance*, March 6, 1967.
218. "Bishop to Dedicate Notre Dame Academy Building," *Staten Island Advance*, June 5, 1963.
219. CSI, "History," https://csitoday.com.
220. "S.I. College Slated for $1.4 Mililon [*sic*]," *Staten Island Advance*, December 7, 1965.
221. "Borough Colleges Improve Physically, Academically," *Staten Island Advance*, November 20, 1964.

Chapter 14

222. Edward Ranzal, "12-Area 'City' of 300,000 Proposed for Staten Island," *New York Times*, May 14, 1970.
223. Regional Plan Association, "A Re-Appraisal of Staten Island," 1964.
224. "Urban Renewal," accessed at dictionary.com on January 3, 2024.
225. Roberts, "Development of Staten Island."
226. "Realty Board Opposes Renewal Project," *Staten Island Advance*, February 8, 1963.
227. "Annadale Renewal Plan Is Opposed by Home Builders," *Staten Island Advance*, February 14, 1963.
228. "Urban Renewal," *Staten Island Advance*, February 16, 1962.
229. S.S. McSheehy, "Foes of Urban Renewal Demand Plan Be Killed," *Staten Island Advance*, September 10, 1963.
230. McSheehy, "Foes of Urban Renewal."
231. "Land Now Worth $50,000 an Acre," *Staten Island Advance*, November 1965.
232. Steven Roberts, "New Town' Urged on Staten Island," *New York Times*, November 9, 1967.
233. Joseph Fried, "City Offers a 'Model' Plan for S.I. Area," *New York Times*, January 22, 1969.
234. Ranzal, "12-Area 'City.'"
235. William Huus, "Private, Public Cooperation Called Key to South Richmond," *Staten Island Advance*, February 24, 1971.
236. Huus, "Private, Public Cooperation."
237. William Huus, "Pre-servicing 'Was a Key Element in South Richmond Plan," *Staten Island Advance*, February 22, 1971.
238. Huus, "Pre-servicing."
239. James Whitford, "Reston Developer Lists Plans for Island Growth," *Staten Island Advance*, November 15, 1966.

240. Whitford, "Reston Developer Lists Plans."
241. Edward C. Burks "City Hall Dominance of S.I. Plan Is Fought," *New York Times*, April 27, 1971.
242. Glenn Fowler, "A Village of 2,000 Town Houses Will Rise in S.I.'s Arden Heights," *New York Times*, March 5, 1970.
243. Fowler, "Village of 2,000."
244. "Progress Fells Trees," *Staten Island Advance*, May 22, 1965.
245. "B.P. Seeks to Keep Roads of Country Club 'As Is,'" *Staten Island Advance*, January 29, 1968.

Chapter 15

246. "Cars, Crocuses Crowd S.I.," *New York Times*, April 10, 1964.
247. Gay Talese, "Down to the Sea for a Nickel," *New York Times*, July 30, 1961.
248. Joe Goldberg, "Staten Island: New York's Last Frontier," *Today's Living*, August 30, 1959.

Chapter 16

249. George O. Reddy, "Commissioner Promises More Police for Island," *Staten Island Advance*, May 4, 1965.
250. Leonard Novarro, "Burglary Rate Increases, but It's Only Temporary," *Staten Island Advance*, November 7, 1965.
251. "Island Needs Own Jail, Braisted Tells Parley," *Staten Island Advance*, September 17, 1965.
252. Drew Fetherston, "Alarmed by Assaults, Residents Form Patrol," *Staten Island Advance*, March 29, 1965.
253. "Householders Offered Anti-Burglar Guidelines," *Staten Island Advance*, April 3, 1966.
254. Raymond A. Wittek, "For Our Protection: Instant Policemen," *Staten Island Advance*, April 30, 1967.
255. George O. Reddy, "Police Bolster Forces on Island," *Staten Island Advance*, January 22, 1967.
256. "Sharp Rise Reported in Island Crime," *Staten Island Advance*, February 12, 1967.
257. William Tow, "Cops Launch Drive to Get Cars Locked," *Staten Island Advance*, October 5, 1967.

258. Victoria Flanders, "Car Insurance Rate Boost Looms for Island Drivers," *Staten Island Advance*, January 23, 1966.

Chapter 17

259. "Memorial Hospital Plans Expansion to Meet Needs," *Staten Island Advance*, January 16, 1964.
260. Ambrose L. Gillon Jr., "Builds, Adapts to Meet Needs of South Shore," *Staten Island Advance*, April 24, 1965.
261. Sister Miriam Eveline, "St. Vincent's Faces Future with Confidence, Purpose," *Staten Island Advance*, April 24, 1965.
262. Julie Summers and Drew Fetherston, "Island Hospitals Being Pushed to Expansion," *Staten Island Advance*, September 23, 1965.
263. Summers and Fetherston, "Island Hospitals Being Pushed."
264. Alfred G. Haggerty, "State School Opened 6 Years Late, but Soon Caught Up," *Staten Island Advance*, March 24, 1964.
265. Alfred G. Haggerty, "Hundreds Labor to Lessen Hardships for the Handicapped," *Staten Island Advance*, March 23, 1964.
266. Haggerty, "Hundreds Labor."
267. Haggerty, "Hundreds Labor."
268. "Willowbrook Mile," Archives of the College of Staten Island, https://www.csi.cuny.edu.
269. "35 Known Dope Addicts on Island, D.A. Reveals," *Staten Island Advance*, October 20, 1962.

Chapter 18

270. Philip Brittain, "Korvettes Starts a Big, Smooth Rush of Shoppers," *Staten Island Advance*, September 25, 1964.
271. Brittain, "Korvettes Starts."
272. "Rosy Retail Future Painted by Macy Exec," *Staten Island Advance*, November 1, 1966.
273. John J. O'Connor, "Ranks of Island's Fruit Peddlers Have Dwindled to 3," *Staten Island Advance*, June 6, 1965.

Chapter 19

274. "Crucial Decade for Richmondtown," *Staten Island Advance*, May 10, 1970.

Chapter 20

275. "History," Church of Our Lady of Pity, accessed via the Internet Archive, https://web.archive.org/web/20171215110714/http:/churchofourladyofpity.com/about/history.
276. "Staten Island Expressway," NYCRoads.com, http://www.nycroads.com.
277. James Ferreri, "The Time: Circa 1965," Restory (Facebook group), June 12, 2022, https://www.facebook.com/groups/1058244067995638.

Chapter 21

278. John J. O'Connor, "After Their Battle Is Won, What Do Civic Groups Do?" *Staten Island Advance*, May 8, 1966.
279. William O'Connell, "Civic Group Builds Up Fight Against Pipeline," *Staten Island Advance*, April 2, 1965.
280. "Construction of Huge Gas Tank Scheduled," *Staten Island Advance*, January 18, 1967.
281. "33 Cases Settled."
282. "33 Cases Settled."
283. "33 Cases Settled."
284. Raymond Wittek, "D.A. Asks Mayor to Halt Tanks," *Staten Island Advance*, February 17, 1973.
285. Lawrence De Maria, "Protection Issue Smolders, but Village Is Heartened," *Staten Island Advance*, December 26, 1971.

Chapter 22

286. S.S. McSheehy, "Water Tunnel Hailed as Answer to the Future," *Staten Island Advance*, February 21, 1962.
287. "Water Tunnel Rock Put to Work in Many Areas," *Staten Island Advance*, June 11, 1967.

288. Raymond C. Fingado, "Phone Company Thinks Ahead," *Staten Island Advance*, April 30, 1960.
289. "Utility Firms Enjoy Year of Expansion," *Staten Island Advance*, February 2, 1969.
290. "Con Edison Allots 2 Million for New Work on Island," *Staten Island Advance*, February 3, 1960.
291. Cy Berlowitz, "Producing Plenty of Power for Lighting and Heating," *Staten Island Advance*, October 13, 1965.
292. "Con Ed Plans $68 Million Travis Plant," *Staten Island Advance*, January 7, 1966.
293. "Con Ed Defuses Plan for Bay Atom Station," *Staten Island Advance*, December 12, 1967.
294. "South Shore Atomic Plant Still Possible," *Staten Island Advance*, May 25, 1968.
295. "Con Ed Offers Buried Wiring," *Staten Island Advance*, January 11, 1968.
296. "All Electric Homes Get Con Ed Bonus," *Staten Island Advance*, January 18, 1968.
297. John Heyke, "Gas Preferred in New Homes," *Staten Island Advance*, April 30, 1960.

Chapter 23

298. S.S. McSheehy, "Maniscalco Accepts His Defeat in Stride," *Staten Island Advance*, November 4, 1965.
299. Douglas Martin, "Albert v. Maniscalco, 90, a Staten Island Leader," *New York Times*, September 25, 1998.
300. Alan S. Oser, "Housing Patterns on SI Resist Change," *New York Times*, October 31, 1971.
301. "'Head Acre' a Nice Headache," *Staten Island Advance*, January 25, 1969.

In Conclusion

302. "Population Now Figured at 300,000," *Staten Island Advance*, March 5, 1967.
303. Murray Schumach, "Verrazano Bridge Casts Long Shadow Over Staten Island," *New York Times*, June 28, 1969.
304. Michael Azzara, "Master Plan for Island Unveiled," *Staten Island Advance*, March 1, 1970.
305. Azzara, "Master Plan."

306. John Corry, "Staten Island: The City's Offshore Bucolic Niche," *New York Times*, July 18, 1972.
307. "Gypsum Co. Officials Neighbors Meet with D.A. on Complaints," *Staten Island Advance*, October 8, 1965.
308. "Total Population."
309. "New York City's Population Estimates and Trends," NYC Department of City Planning Population Division, https://s-media.nyc.gov.
310. Michael Dominowski, "Lost Highway," *Staten Island Advance*, April 29, 2012.

ABOUT THE AUTHOR

Patricia M. Salmon was born in Brooklyn. She is a retired curator of history from the Staten Island Museum. She is the author of five books on Staten Island history: *Staten Island's Brewery Barons*, *Murder & Mayhem on Staten Island*, *Realms of History: The Cemeteries of Staten Island*, *Staten Island: Murderers and Mysteries of the Forgotten Borough*, and *The Staten Island Ferry: A History*. Ms. Salmon is currently working on a book about powerful women from Staten Island's past. She is the official historian of the Friends of Olmsted-Beil House.